Alok Jha is a journalist and broadcaster based in London. He is science correspondent for ITN and, before that, was science correspondent at the *Guardian*. He has presented science programmes for BBC2 and BBC Radio 4. Alok received a science-writing award from the American Institute of Physics in 2014, was named European Science Writer of the year in 2008, and has been shortlisted for awards by the Association of British Science Writers and the Medical Journalists Association.

ALOK JHA

THE WATER BOOK

headline

First published in 2015 by
HEADLINE PUBLISHING GROUP

First published in paperback in 2016 by
HEADLINE PUBLISHING GROUP

1

Cataloguing in Publication Data is available from the British Library

Paperback ISBN 978 1 4722 0953 5

Typeset in Miller by Palimpsest Book Production Ltd, Falkirk, Stirlingshire

Printed and bound in Great Britain by Clays Ltd, St Ives plc

MIX
Paper from
responsible sources
FSC® C104740

Headline's policy is to use papers that are natural, renewable and recyclable
products and made from wood grown in sustainable forests. The logging
and manufacturing processes are expected to conform to the environmental
regulations of the country of origin.

HEADLINE PUBLISHING GROUP
An Hachette UK Company
Carmelite House
50 Victoria Embankment
London EC4Y 0DZ

www.headline.co.uk
www.hachette.co.uk

To my parents

Contents

Aldershot Library 4
RENEW ONLINE at www.hants.gov.uk/library or
phone 0300 555 1387

LOVE YOUR LIBRARY

**** Summer Reading Challenge – Gadgeteers
****

Saturday 16 July – Saturday 17 September

Sign up at your local library from 16 July and join
the Gadgeteers to discover the amazing science
and innovation behind the world around you!

****

Customer ID: ******0803

Items borrowed today

Title: In the dark
ID: C016089759
Due: 30 August 2022

Title: Knife edge
ID: C016890374
Due: 30 August 2022

Title: The water book
ID: C016222012
Due: 30 August 2022

Total items: 3
Account balance £0.00
01/08/2022 12:45
Items borrowed: 3
Overdue items: 0
Reservations: 0
Reservations for collection: 0

Download the Spydus Mobile App to control your
loans and reservations from your smartphone.
Thank you for using the Library.

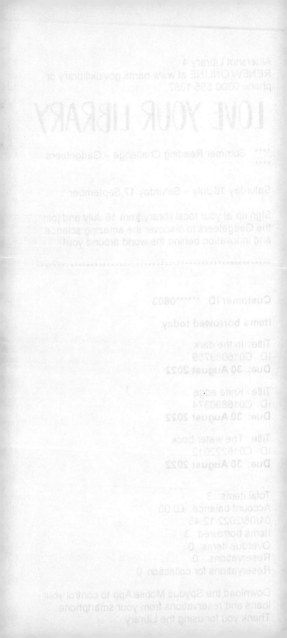

Introduction

'If I were called in
To construct a religion
I should make use of water'

'Water' (1954), *The Whitsun Weddings*,
Philip Larkin

It's raining outside. If not where you are, then somewhere on the Earth at this very moment, water is falling from the sky. It might be droplets or snowflakes, sleet or hail. Water is always moving – under your feet in unseen aquifers and in the pipes laid down by engineers to move food and waste around our cities. It moves next to you in trees and plants, sucked from the ground to feed their leaves. Water solidified the concrete of the walls around you or produced the wood or plastic for your chair, the paint on the walls and the drink by your side. And you might hear it nearby, in the sea, a river or a lake.

It works inside you, a thick treacle that looks unlike any other water you have ever encountered. It moves around in your blood (it *is* your blood), keeps your proteins and DNA working and in their correct shapes and transports nutrients and signals in and out of cells. Each living cell is mostly water, each one differentiated only a fraction from purity by a few chemicals.

To humans, though, water is more than a mere chemical, and more than a functional ingredient for life. In fact, we rarely think of it as either of these things specifically. Instead, thinking of water immediately brings to mind a cultural object, constructed from the overlapping stories of hunters, poets, Olympic swimmers, factory-workers, novelists, ecologists, water engineers, farmers, consumers, chemists, historians, theologians, divers and astrobiologists. Each will give you a different view. All will be correct. Put them together and you still have an incomplete picture.

How can something so common and familiar be so difficult to describe? What we see when we look at water depends on the time frame in which we see it, of course. In our personal encounters with water, it is infinitely yielding. But over the course of centuries, it writes its impulses indelibly on the landscape. 'There is nothing softer and weaker than water,' writes Lao Tzu in the *Tao Te Ching*. 'And yet there is nothing better for attacking hard and strong things.'

Water nourishes and soothes us. But this same stuff also carved the Grand Canyon out of solid rock over the course of millennia, and every day thunders down with unimaginable fury at Niagara and Victoria Falls.

In the tsunami that flowed across the Indian Ocean in 2004, water was the medium that expressed a tension in the Earth's crust that had been gathering in force and latent energy over the course of thousands of years, killed hundreds of thousands of people, and wrought devastation for millions more.

This liquid, a substance of life, is also one of fear. Roiling waters can take us away from air, engulf us and disorient us. Though we need and crave it, water can be a tantalising poison for thirsty sailors. Its paradoxical nature can be nightmarish, as Coleridge knew: 'Water, water, every-where / Nor any drop to drink.'

Novelists, poets and journalists have talked of the fore-boding of a body of water, the dark unknowability of the sea, the loss when something slips overboard and to oblivion beneath the surface. 'Consider the subtleness of the sea,' wrote Herman Melville in *Moby Dick*. 'How its most dreaded creatures glide under water, unapparent for the most part, and treacherously hidden beneath the loveliest tints of azure. Consider also the devilish brilliance and beauty of many of its most remorseless tribes, as the dainty embellished shape of many species of sharks. Consider, once more, the universal cannibalism of the sea; all whose creatures prey upon each other, carrying on eternal war since the world began.' Better to remain safe on the verdant land, he counselled, and out of the 'appalling' ocean that surrounded it.

We know more about space than we do about the furthest reaches of the oceans because, despite the difficulty and expense of escaping the atmosphere, it is actually easier and less dangerous than dealing with the crushing pressures of the deep sea.

Still, water is the life giver and no known life exists without it. This chemical has been our key to exploration as we look for life among the stars. In this search, we have been looking for worlds like our own, a snapshot of the primordial Earth, perhaps, as it might have been before we evolved to colonise it, changing it slowly beyond recognition.

Water courses through us, our societies and our planet. But look at it rationally and this is a profoundly strange chemical that bends and flexes the usual rules of chemistry: why does ice float on water? How can liquid water store so much more heat than anything else? How does it manage to so carefully choreograph the behaviour of so many biolog-ical molecules inside our cells? Why is water not a gas at room temperature, given how light its molecules are? All of these things, just a few of dozens of anomalies and complexities that mark water out as a strange chemical, have been critical to the formation and evolution of complex life. If water behaved like everything else, the Earth would look very different and none of us would be here to know about it. Given this fundamental importance to our world and to our biology, it is perhaps surprising that we have only recently begun to understand why water behaves the way it does.

At some level, well before we began to understand its strange chemistry, we have always known there was something special about water. It is the only natural substance in which we have invested so much culture and holiness.

One of the roots for the word 'water' comes from the Sanskrit 'apah', meaning 'animate', something that gives life. It has survived, via the Old Latin word for river, 'abnis', in the modern Irish and Scots Gaelic words for river, which are 'abhann' and 'abhainn' respectively. In many Western languages, the Proto-Indo-European roots 'wodor/wedor' sit behind the words for water itself – turning into the familiar 'water' in English, 'wasser' in German, Icelandic 'vatn' and Russian 'voda'. The same root words give us 'wet' and 'wash'.

Along with fire, earth and air, water is one of the four classical elements, and it was associated with phlegm as one of the body's humours. In the East, water was one of the five elements of Chinese Taoism with earth, fire, wood and metal.

Civilisations do not spring up far from sources of water. The greatest human settlements have coalesced around rivers and seas, the water bringing them life and trade. Mesopotamia (from the ancient Greek for 'land between two rivers') once flourished between the Tigris and the Euphrates. Egyptians (ancient and modern) have depended on the Nile. Modern mega-cities such as London, New York, Tokyo, Hong Kong and Mumbai rely on their easy access to water.

Cultures have marked their greatness by their mastery over water. The Chinese symbol for 'political order' is made from the characters for 'river' and 'dyke' and the meaning is clear: whoever controls water controls society. Control of water allowed empires to grow to hitherto-unprecedented scales – the Romans were able to found a new type of civilisation after they built their first aqueducts, more than 2,000 years ago. Today, the Three Gorges Dam across the Yangtze River is the biggest hydroelectric project in the world, a sign of the progress of an emerging technological and economic superpower. Gigantic dams, intricate canals and cutting-edge water-processing technologies across the world – from the United States to India to Chile – are all signs of a country's determined march towards controlling its own future.

The human relationship to water is complex, multi-dimensional and, like a fractal, infinitely more intricate the closer you look. A single book could never hope to do justice to all of it. Instead, by wandering historical, physical and scientific waterscapes, I want to bring together threads of the human story of water that might, at first sight, seem disparate and unconnected.

Those threads will take into account water at all scales – we begin surrounded by water, on an expedition to see the ice fields of Antarctica; consider the feeling of moving on its surface and take stock of the Earth's entire physical stock of water, in all of its frozen, gushing and wispy forms. We will look at how water moves, how that movement creates our weather and moved ships to new lands. At the smallest scale, we will look at how individual molecules of water came to be and then how groups of them behave in ways that still baffle scientists but which have become the *sine qua non* for life (and the culture and society that goes along with our life). Eventually our story will leave our home planet – just like our water will one day – to consider what else there might be out there, carved by water, among the stars.

The author Tom Robbins said that humans were invented by water as a way to transport itself; I would go further – water invented us as a way to appreciate itself. This is a story that will connect you, via one strange molecule, to everyone and everything else and the rest of the universe.

> 'And I should raise in the east
> A glass of water
> Where any-angled light
> Would congregate endlessly.'

'Water' (1954), *The Whitsun Weddings*,
Philip Larkin

Part I

HYDROSPHERE

ONE

Departures

'Whenever I find myself growing grim about the mouth;
whenever it is a damp, drizzly November in my soul; whenever
I find myself involuntarily pausing before coffin ware-
houses, and bringing up the rear of every funeral I
meet . . . then, I account it high time to get to sea as soon
as I can.'

Herman Melville, *Moby Dick*

We left the southernmost tip of New Zealand on a cold,
bright afternoon. Six months in the planning, this
was the trip of a lifetime: a month sailing to, around
and back from the most remote part of the most remote continent on Earth, Antarctica. A metaphor for isolation and loneli-
ness as much as a location in itself, a place that is still distant in
our hyper-connected world. A century ago, the journey we were
about to start was the exclusive preserve of teams of men led by
heroes whose names became etched into history – Scott, Mawson,
Shackleton and Amundsen – men who had returned (if they
returned) with stories of desolation, drama and tragedy. The awe
of discovering a new land and surviving its bitter, inhuman
environment had captivated civilised society a century ago.

The continent is now more accessible but still not easy
to get to, its visitors limited mainly to small bands of

scientists from around the world, groups of scholars and explorers who make annual trips to Antarctica. They come to better understand the Earth's weather and climate systems, by examining its most pristine and untouched part, or to look up at the clear skies in order to stare deep into the history of the universe. The only other visitors, during the summer months of October to February, are VIPs, supply-ship crews and well-heeled tourists, the last of whom can only look at the coastline from a safe distance, occasionally venturing briefly onto the ice when conditions allow.

My expedition to this place, The Ice as I later learned to call it, was part of a private science expedition following in the footsteps of the great British–Australian explorer and geologist Douglas Mawson. He had first ventured to the continent with Ernest Shackleton in 1908 and then, before he was even thirty years old, managed to raise the money to lead his own expedition in 1912. He was an innovator – the first man to take an aeroplane to Antarctica and the first to set up and use wireless communications to send information back home. He sailed under the flag of Australia and much of that country's modern claim to the frozen continent can be traced back to Mawson's expeditions of 1912 and 1929.

Our ship's horn blasted a long monotone at 2 p.m. as we moved backwards from the dock at Port Bluff, New Zealand. Behind us the harbour continued for a few miles between two strips of green land. A cliff rose up in the distance to our starboard side, a marker of the start of the Foveaux Strait. Beyond that and over the horizon lay the wilds of the great Southern Ocean.

As soon as the ship had raised its anchor, it began to move simultaneously in all three dimensions. It was a subtle thing, a feeling in the gut rather than anything I could see, but I could sense that the hard, sure ground beneath my feet had become a little unsteady. I had a mild feeling of discomfort and a slight difficulty in keeping balanced. Pushing the strangeness to the back of my mind I walked onto the observation deck of the ship, positioned directly above the bridge, where several dozen members of the expedition had lined up around the curved bow. We swapped names for the first time, shook hands, cheered and took pictures. Some people furiously tapped out text messages on their mobile phones in an attempt to send a final signal over the airwaves, using the very last bars of reception that any of us would get for the next month, to say goodbye to friends and loved ones.

By the time the ship passed the cliffs, we had been rocking for a good half an hour. Land was still reassuringly close behind us but, ahead, it was endless blue. Already I felt small on this ship.

And then came the nausea.

Seasickness is nothing you can control. Well before I had embarked on the expedition, before I had come within a hundred miles of the ship, I had been told by doctors that the sickness can take hold even in the most experienced sailors, rendering them unable to move, incapable of eating and in constant fear of vomiting. Who would suffer, how bad it would get and how well those individuals would manage to deal with the discomfort were all unknowns.

Douglas Mawson – one of the hardiest of the great polar explorers of the Heroic Age – had himself been struck with sickness almost immediately after leaving port in Hobart, Tasmania. This is a man who would later watch two of his closest friends die, one from madness and poisoning and the other lost to a crevasse, during one of the most epic and tragic stories in the pantheon of great stories that are told about Antarctica. This is the man who spent a month trudging hundreds of miles across Antarctic tundra back to his base at Commonwealth Bay, by himself. Mawson ran out of food and had to eat his sledging dogs; he had to re-attach the soles of his own feet with lanolin every day as he trekked through the icy wastes; and, when he did get back to the rest of his team at the continent's coast, he just missed his ship home. This man, made of steel, was struck low on the way to the frozen continent by nausea. For a week, he didn't emerge from his cabin aboard his ship, the *Aurora*, and his otherwise meticulous diaries are empty for his first few days on the Southern Ocean.

Modern sailors (experienced or amateur) can take drugs to ward off the effects of the sickness. Different drugs do different things – some help with balance while others reduce the discomfort of the vomiting cramps and headaches. The best ones just make you drowsy, a useful tactic that prevents you moving around too much while unwell and eventually brings on the only thing that actually works in the battle against nausea – blessed unconsciousness. None of these drugs seem to work very well at getting you used to seasickness, which is something that your inner ear, brain and legs have to learn to deal with all by themselves.

The first evening, after an hour on deck watching land slip further away and trying to focus on the adventure ahead, I could no longer ignore the rising throb above my eyes. I made my way down a few flights of stairs and stumbled onto my bunk. By now the ship was jerking up and down, left and right. It felt entirely random but my brain searched out rhythms nevertheless, which I would latch onto in a vain attempt to predict and control the movement and try to push it to the background. Any regularity, real or imagined, never lasted long, though. Eventually I found myself thinking that I could probably get used to the movement if it would just stop for one minute so that I could catch my breath and let the throbbing around my eyes settle down. A single minute would be fine. I could get my bearings, catch my breath and steel myself for another onslaught. Just one hour into my month-long journey, I wanted a break from the ship and was counting the days until I could get off.

But I knew all of this thinking and wishing was futile. There was no way off until the ship landed at the Antarctic coast in around ten days and I just had to get used to the movement.

Less than two hours from port, I lay on my bunk,

curtains drawn around me and eyes closed, at the mercy of an unseen, rolling sea. Excitement about the expedition had carried me to this moment. I was prepared for the landscapes, the meanings and history of where I was about to go. But, in taking up all my thoughts, that excitement had excluded any contemplation of the visceral logistics, the physical feeling of the journey.

It filled my mind now, though. Everything elsewhere had faded and my mind was zeroed in on this moment.

Several hours before the ship had left port, I had stuck a patch behind my ear, as recommended by a pharmacist in Invercargill, where I had been staying for the past week making final preparations for the trip south. A lot of expeditions to the Antarctic left from the port near this town and the pharmacist stocked many different types of pills and remedies to ward off nausea on the Southern Ocean. Which one would work best would be a process of trial and error but his main recommendation had been that preventing seasickness was far better than trying to cure it after it had taken hold.

The skin patch, infused with medication, was meant to work for up to three days, guarding against sickness. But several hours after I had stuck it on, and with land now out of sight, the damned thing was having no effect. There was no way off this ship and it was hopeless to think the vibrations going through my skull might just stop at any moment.

In this fog, I remembered that I had back-up seasickness pills. The pharmacist had told me they were meant for sparing use, since they induced drowsiness and were therefore not too useful if you wanted to do any work. Right now, work was the furthest thing from my mind and instead I focused on trying to remember where I had packed the medicine.

Retrieving the pills became a carefully planned operation. The movement of the ship meant that standing up for more than half a minute was impossible without an overwhelming need to either vomit or go and lie down again. I had to organise my retrieval mission meticulously to fit into my allotted 30-second time slot. I mentally went over it a few times as I lay on the bunk. The pills were in the first-aid kit at the bottom of my main suitcase. The suitcase was under my bunk. The sink was about a metre away.

Sit up on the bed, slide the suitcase out, unzip it, pull out the first-aid kit, take the blister pack of tablets, push two of them out and straight in my mouth, stand up, turn to the sink, draw a cup of water, drink it all down and stumble back onto my bunk.

I went over this several times in my head before taking a few deep breaths and swinging my legs over the side of the bunk and sitting up straight. Even that mild action gave me a head rush that forced me to close my eyes and take a few moments' pause before moving on to the next part of the plan: the suitcase.

Holding onto the bunk with one hand, I reached down underneath, found the handle of the suitcase and yanked at it. It wouldn't budge. I yanked again. It was caught on something. The lateral rocking of the ship for the past few hours had shifted the suitcase around under the bed and it had managed to move itself into a position in which it was stuck fast. My plan hadn't taken account of this.

I would have to look under the bunk to try to free the suitcase and, crucially, my seasickness pills. I steeled myself, stood up, spun around and kneeled on the floor. Looking underneath the bunk, I saw that the suitcase had manoeuvred itself behind a set of drawers attached to the underside of the bed. I pulled it free and quickly unzipped it, all the time my brain counting down the seconds until it would force me to throw up. I located my first-aid bag and a worrying thought immediately came over me. To save space in the bag, I had thrown away all of the boxes that my medicines had come in. I began to worry that there was no way I could distinguish the seasickness medication from the painkillers, antibiotics and antihistamines that I knew were also in there. And, anyway, in the dark of the cabin how would I focus on the tiny words printed on the foil? Again I hadn't factored this delay into my commando plan to retrieve my medication.

I ripped open the first-aid bag and rifled through the dozen or so identical-looking silver blister packs. I was in luck – the seasickness pills, Stugeron, were printed with their name in huge blue type across the back of the pack, almost as if the makers had pre-empted this situation, knowing that their tablets would need to be deployed in conditions when focusing on reading labels might be difficult. I popped out two of the small white tablets and dropped the rest back on the floor – being tidy wasn't a priority right now.

By then, though, the nausea had also caught up with me and I was forced to lie down on the floor next to my open suitcase, still clutching the two tablets in my right hand. I closed my eyes and tried to ignore my brain's command to go and find a place to vomit.

I don't know how long I lay there, waiting for my head to stop spinning. Eventually I summoned the strength to slide over to the foot of the sink, slowly prop myself up against it, fill a metal cup with water and bring it up to my mouth. I dared not look at myself in the mirror above the sink. I swallowed the tablets and gulped down a few sips of water after them. Leaving the cup on the hook next to the mirror, I walked the two steps back to my bunk and collapsed.

These backup seasickness pills, I had been told, were for emergencies, when we were far into the journey across the Southern Ocean where the seas were notoriously rough. A few hours off the coast of New Zealand, where the seas promised to be about as gentle as they would ever get on our expedition, I had given in. My eyes dropped soon after, just as I heard the announcement, over the ship's loudspeakers, for dinner. Who on Earth, I remember thinking as I drifted off, would want to eat anything now?

Nausea is the brain's response to sudden unfamiliarity, a way to stop you from doing whatever new unknown thing you're attempting that might be causing you problems. When the brain suspects something is wrong, when its senses don't correspond with each other and there has been no time for adjustment, it will wrest control of your body in the only way it knows how. There is always a nausea associated with leaving behind the known – a stable ground beneath your feet, for example. That unease is multiplied when the most familiar thing turns out to be shocking and strange. Our subject – water – is one of those things.

Water is the only substance on Earth whose chemical formula has entered the vernacular. The simplicity of the 'H_2O', however, belies the complexity of its chemistry. 'Of all known liquids,' wrote the chemist Felix Franks, 'water is probably the most studied and least understood.'

If that sounds strange right now, then you are in good company. Water is so common, so familiar as to seem boring. Thinking about it too deeply might seem pointless, a waste of time better devoted to tackling other mysteries. But a shadow lurks behind this substance, a shifting darkness that baffles and confuses anyone who dares to peer at it for too long. 'There aren't terribly many scientists who study liquids,' wrote scientist and author Philip Ball, 'but I was once one of them, and we tended to shun water. This might make us sound like bakers with an aversion to bread, but we had a good reason: water broke all the rules. There is a perfectly good "theory of liquids" that has been painstakingly developed since the late nineteenth century, and it is astonishing what it can accomplish in terms of explaining what liquids are and what they do. But it is of rather little use for understanding water.'

That strangeness is visible (though you may not see it as such) every time you drop an ice cube into a drink. It's hard to fathom anything wrong with that simply because we're so used to seeing it, but think about it for a moment. In front of you is a solid, floating on its liquid. Solid wax doesn't float on melted wax; solid butter doesn't float on its melted form in a hot saucepan; rocks don't float above lava when it spews out of a volcano. When a liquid cools and starts to becomes solid, it contracts in size and the solid parts should fall to the bottom of any liquid that is present.

Instead, ice floats because water does something strange when it freezes – it expands. Anyone who has ever tried to rapidly cool a bottle of champagne by leaving it in a freezer overnight will quickly confirm that this expansion is a powerful force, strong enough to shatter glass. Freezing water can also split water pipes during cold winters; water that seeps into cracks in buildings can freeze and break walls. It seems like a small and inconsequential curiosity. But this anomaly of water, one of a collection of strange and unique behaviours that build into an overall refusal by this substance to conform to the normal rules of liquids, has shaped our planet and the life that exists on it.

Through aeons of cycles of freezing and melting, water has seeped into giant boulders, cracked those rocks apart and broken them up into soil. Ice floats in our drinks but also across our oceans, rivers and lakes. Sea ice and glittering icebergs are familiar around the polar seas, as are the still, frozen lakes, nearer to where we live, in winter. In those frozen lakes and rivers, the ice does more than decorate the surface; it insulates the water underneath, keeping it a few degrees above freezing point – and, crucially, liquid – even in the harshest of winters. Water is at its most dense at 4°C and any liquid at this temperature will sink to the bottom of a lake or river. Because bodies of water freeze from the top down (rather than the other way around), fish, plants and other life in the water will almost always have somewhere to survive successive winters, and be able to grow in size and number. Over geological time, this curious anomaly of water has allowed complex life to survive and evolve despite the Earth's successive ice ages – harsh periods when fragile life forms would have otherwise been wiped out on the desiccated, frozen ground and, if water behaved like a normal liquid, in solid seas.

Floating ice is just the start of the strange behaviour of water. In 1963, when Erasto B. Mpemba was in grade 3 at the Magamba Secondary School in Tanzania, he noticed something profoundly weird about water that no-one can explain to this day. He and the other boys at his school would often make ice cream by boiling milk, mixing it with sugar and putting the mixture into a freezer once it had cooled to room temperature. Freezer space was always at a premium, however, and there was often a rush among the boys to get their ice cream mix into the freezer. One day when he was boiling up his milk, Mpemba noticed that another boy was hurrying to get his own mixture into the freezer first by leaving his milk cold. Mpemba knew that if he waited for his ice cream mixture to cool before he took it to the freezer, there would be no space left. So he risked damage to the refrigerator and put his hot ice cream mixture into the machine, alongside his friends' cold mixture. An hour and a half later, the boys went back to check on their ice creams and they found something peculiar: while Mpemba's mixture had frozen into ice cream, the other boy's mixture was still only a thick liquid.

The Mpemba Effect, as his discovery is known, is the observation that hot water freezes faster than cold water when both are placed in the same sub-zero conditions. Though ridiculed by his teacher, Mpemba was not alone in noticing this peculiar effect of water – Aristotle, Francis Bacon and

René Descartes have all written about it. In investigating his own findings further, Mpemba saw that local ice-cream sellers in the nearby town of Tanga also made their products in the same way that the young Mpemba had accidentally done at school – by placing hot milk straight into the freezer – because they knew it froze faster that way.

There aren't any definitive answers about what causes the Mpemba Effect. It could be because of the evaporation taking place at the surface of the hot milk, removing heat as well as some water so that the resulting volume is smaller and colder and, therefore, freezes faster. Another idea is that hotter milk contains fewer dissolved gases, which typically act as impurities and lower the liquid's freezing point.

A third possible explanation involves supercooling. If you cool water to its commonly accepted freezing point (0°C at a pressure equivalent to one atmosphere), it will form ice only if there are places in the liquid where a crystal can start to get an initial purchase and start to grow. These starting points could be bubbles, for example, or rough areas on the surface of the container. If there are none of these around and the temperature keeps dropping, the water can become supercooled and remain liquid well below its expected freezing point. Supercooled water is not entirely stable though, and will instantly freeze if nucleation is somehow started within it. (It works the other way too – superheated water can remain liquid well past the usual boiling point of 100°C on the surface of the Earth).

Since hot water has fewer impurities and potential nucleation points than colder water, it can become supercooled more easily. As colder water freezes, it forms a layer of ice at the top that insulates the rest of the volume underneath from the cold air outside. That slows down the rate at which the rest of the liquid can freeze. In the case of supercooled water, however, there is no such ice and no insulation. The liquid therefore freezes more rapidly once nucleation does somehow get started.

With the Mpemba effect, we're getting a window into the confusing ways in which water reacts to changes in temperature. In fact, here lies one of the strangest anomalies of all – that water is a liquid at all at the temperatures and pressures on the surface of the Earth.

Given that water molecules are made from two such light atoms, the textbook rules of chemistry say that our planet should have no liquid oceans. All of the water on our planet should rightly exist as a vapour, part of a thick, muggy atmosphere that sits above a bone-dry surface. Hydrogen sulphide (H_2S) is a gas, even though it is twice the molecular weight of water. Ammonia (NH_3) and hydrogen chloride (HCl) are gases despite being similar-sized molecules to the more famous H_2O.

Many of the anomalous behaviours we have encountered so far can be traced back to an intriguing ability that water molecules have to temporarily attach themselves to things, particularly each other. This 'stickiness' not only explains why ice floats and water behaves in odd ways with regards temperature, it is the key to understanding why water became the medium of life, over and above anything else.

At the macroscopic level, water's stickiness gives it a strong surface tension, a property that insects use to skate across the surface of ponds, their tiny masses light enough to avoid breaking through the surface of the water molecules bound together at the surface. That same stickiness means that water molecules easily pull each other up through empty pores and vessels by capillary action – the ability of a liquid to flow in very narrow channels without the action of (and often in opposition to) an external force, such as gravity. This is how water moves up so easily through a tissue paper or a sponge; it is also how plants suck water up from deep below the surface of the Earth in order to nourish the leaves and branches growing in the sunshine.

On our everyday level, water's stickiness is profoundly useful to us because it makes liquid water so incompressible. Squeezing orange juice out of a carton, for example, or pumping water around plumbing in a house or through a hose to put out a fire or water a garden can only work because applying a pressure at one side of a volume of liquid water moves it so easily. If water was more easily compressed (or, rather, if it was as easily compressed as other liquids) moving it around in the industrial quantities we need for modern life would be expensive and difficult. (A liquid being incompressible might not sound so abnormal, by the way, but water takes it to a different and weird level – even at a mile deep, the ocean's water is only squashed in volume by only about 1 per cent.)

All of this stickiness comes about because, at the molecular level, water molecules are attracted to each other to an extent we do not see with other liquids. Each water molecule is shaped like a shallow V, with the large oxygen atom at the apex and the smaller hydrogen atoms at the ends of the arms. The atoms within an individual molecule are bound together by covalent bonds, a type of chemical bond that is at the heart of a great number of chemicals in the universe, in which the atoms share their outer shells of electrons. Overall, each H_2O molecule is electrically neutral but zoom in and you find that the two hydrogen atoms are fractionally positive, compared to the fractionally negative oxygen atom at the centre of the V – the molecule has positive and negative electrical sections, like north and south poles on a magnet. And like magnets, this polarity in water leads to attraction between different molecules as the hydrogen atoms of one water molecule are weakly drawn towards the oxygen atom of another molecule. This 'hydrogen bond' between water molecules is minuscule compared to the covalent bonds within the molecules themselves but its consequences are profound for how water behaves at the temperatures and pressures we are familiar with on Earth and, as we will see, it is critical for the functions of life.

Water molecules can form a total of four hydrogen bonds each with their surrounding molecules, creating a loosely-connected structure that looks like a pyramid where each of the four faces is made from an equilateral triangle. These structures repeat throughout the liquid to create a vast 3D network, with a water molecule sitting at each corner of each pyramid.

But the networks are not fixed. Rather, the water molecules continually dance around each other in three dimensions, bonds between them continually breaking and reforming in trillionths of a second, the pyramid structures forming, breaking and reforming. Those minuscule, ephemeral, ever-shifting molecule-to-molecule attractions add up, however. They give liquid water a little more cohesiveness than is found in other liquids, just enough so that liquid water needs more energy, overall, in order to boil up into a gas. That means we end up with a liquid at the temperatures and pressures on Earth where we might otherwise have expected a gas.

This hydrogen-bonded network also explains why ice is less dense than cold liquid water. In other liquids, where hydrogen bonds do not form networks, the molecules can end up drifting extremely close to each other, meaning the liquid ends up becoming relatively dense.

In liquid water, there are two opposing forces at play. For a start, the hydrogen bonds want to force the water molecules into pyramid-shaped networks that end up leaving lots of empty space between the molecules (they create an 'open' structure). Working against that imposition of order is the inevitable chaotic maelstrom of a liquid. Because the water molecules all tumble around through the liquid, the perfect pyramids of the hydrogen-bonded networks do not always come together. In many places, the water molecules will drift closer than they should if the hydrogen bonds were in total control. On average, those small imperfections in the network of pyramids means that liquid water is not as open as it should be – there is less space between molecules, in other words, than you might otherwise expect if the pyramid structures were always perfectly formed.

When water is frozen into ice, however, those pyramid

structures are forced to form perfectly. The solid network of water molecules in the crystal is therefore forced into its more open, and less dense, state. Hence ice is less dense than water, and it floats on the surface of the liquid.

With ice, we enter another new world of interesting aqueous behavior. The ice we see on Earth, a type known as ice-Ih, is only one way that water can crystallise. At the last count, there were sixteen known phases of ice, each with its own crystal structure, each one forming under different environmental conditions. Only ice-Ih occurs naturally but the other fifteen have been made and studied in laboratories and, who knows, could be our glimpse into the state of water in space and on other worlds.

The polarity of the water molecule has another use fundamental to our life and that of our planet, which is worth mentioning here. The fractional difference in electrical charge across the V-shaped molecule not only makes water molecules stick to each other, it makes water the closest thing we have to a universal solvent, able to tear apart and nudge compounds that have any electrically-charged components.

Sodium chloride (common salt), for example, dissolves in water because the polar water molecules tear the salt crystal apart. Common salt is what is known as an ionic crystal – when they react together, the sodium atoms give up electrons in their outermost shells to chlorine atoms, turning both of them into electrically charged atoms called ions. In the crystal, the positive sodium ions and negative chloride ions are arranged in a regularly spaced lattice.

When water molecules come along, their slightly positive hydrogen ends are attracted to the negative chloride ions and the slightly negative oxygen atom in the water is attracted to the sodium atoms in the crystal. The result is that the crystal lattice breaks into pieces, something we can see with our naked eyes as salt grains dissolve in a bowl of water. The sodium and chloride ions end up floating freely in solution.

Liquid water is such a good solvent, in fact, that it is almost impossible to find naturally occurring pure samples and even producing it in the rarified environment of the laboratory is difficult. Almost every known chemical compound will dissolve in water to a small (but detectable) extent. Related to this, because it will interact with everything, over long periods of time water is also one of the most reactive and corrosive chemicals we know.

This ability to react and dissolve also lies at the heart of water's ability to conduct electricity so well. We all know that it is dangerous to keep electrical wires wet and that you should never go near electrical sockets with wet hands. Pure water, though, is actually a very good insulator because, even though the hydrogen and oxygen atoms are fractionally charged, they cannot move independently of one another in order to set up a current within the liquid. Pure water, though, is hard to find and, as we know in the real world, the liquid will dissolve anything it can find. The electrically charged impurities within water are what make the liquid so conductive. Even clean water exposed to air will dissolve carbon dioxide and turn the gas into ions in solution, which can then transport electrical energy around.

When we reduce water to its constituent parts and begin to take a peek under its hood, we need to occasionally remind ourselves of the importance of what we are dealing with. 'Not just a chemical compound, but a fundamental part of nature,' writes Ball, 'with aspects that are serene, enchanting, enlivening, profound, spiritual and even terrible. In the voice of the babbling stream, says Wordsworth, "is a music of humanity".'

Perhaps it is too simplistic, too reductionist, to say that life on Earth is a consequence of the chemical properties of water. But most of our biological functions (and indeed the movement and presence of water that made our environment so hospitable) can often be brought back to the way water molecules attract and dance around each other.

The properties and anomalies we have started to explore give water its edge in providing the tools necessary to shape the Earth and keep life functioning, and we will explore this in more detail in Chapter Five. Between them, water molecules, their polarity and hydrogen bonds create the mechanisms and scaffolds for life, giving the liquid a unique ability to push and pull basic chemical ingredients around to build the complex, living things we see around us.

Think of a liquid and it will most likely be water. Even if you actively think of something else – blood, beer, apple juice – you're thinking of water with a small amount of other things dissolved or suspended within it. There are other pure liquids that appear in everyday life, such as petroleum or cooking oil, but not many and we hardly think or interact with these liquids to the extent we do with water. Water is so common and so familiar that it is mundane: every day we drink it, touch it, wash with it, wet things, dry things, we boil it, freeze it and even immerse ourselves within it by swimming or bathing in it. The anomalies outlined here are evident to us only because we can, with measurement and objectivity, stand outside our experience to watch it in action and compare it to the other materials we know about. More important, though, is that we live in a world where the environmental conditions allow us to travel easily around so much of the energy landscape of water, around temperatures and pressures where it can slide comfortably between solid, liquid and gas (or sometimes all three at once). We know about this variability, we can see so many nuances and edges to this material, because our world is set up to give it to us and we are evolved to sense it. The more we examine water, the stranger it gets, the more questions appear in front of us. We study it because we are made from it and it is, perhaps, surprising that what we are made from is still such a mystery.

TWO

Southern Ocean

The Southern Ocean is the roughest sea in the world. Spoken about in whispers and glances, it has smashed ships and led sailors to despair for hundreds of years. There is no choice but to cross it, though, if you want to reach Antarctica by sea from any direction. From Bluff to our destination on the east Antarctic coast we would have to sail across 2,400km (1,500 miles) of open ocean, around ten days at sea.

This ocean completely surrounds the Antarctic continent and, between certain southern latitudes, it circles the Earth uninterrupted by land. The winds around those latitudes, the equivalent of the pervasive westerly winds in Europe, were nicknamed the Roaring Forties, Furious Fifties and Screaming Sixties by late nineteenth and early twentieth century whalers, fishermen and sealers who used to hunt in these regions. Because there is no land in their path to dissipate or slow them down, the winds can build up phenomenal speeds as they blow around the planet, gusting up to 80kph on average and whipping up swells of four or five metres on a normal day and far more during storms.

We were alone here, no longer in an environment where there were other vessels within easy reach. In all directions around the ship, hills of water oscillated, joined together and crashed. Even when it was relatively flat, the ocean surface looked like a shattered plate that had been badly glued back together and was pivoting at all the joins at once. Our ship bobbed, insignificant, in this massive ocean.

During storms it was impossible to tell where the sea stopped and sky started, and on many other days the sky was a just a grim, slate-grey strip sitting above the deep blue of the waves. The water, always moving, became terrifying when the ship tilted a little too far to one side and the swell there rose to form a wall of water the height of a skyscraper (or so it felt), but infinitely wider, that rushed towards us.

The stern deck of the ship was relatively low in the water compared to the rest of the ship and, on rougher days, the Southern Ocean would send waves crashing across it. Back there was a storage area of sorts – a few rooms along the side contained scientific equipment, fuel and spare parts for the ship. On the exposed deck, several rubber Zodiacs and amphibious Argo vehicles were lashed to the metal. A crane loomed over one side, used to lift equipment and cargo on and off the ship. The first time I wandered out there during a storm – I don't know why but I wanted to feel the elements, the air and isolation that evening – my legs were drenched as the ship tilted and the deck became submerged in several inches of water.

Our ship was built for these conditions, and worse. A 70-metre long ice-strengthened Russian polar vessel, the *Akademik Shokalskiy* had seen more than thirty years of service in the Arctic and Antarctic regions, initially as a mapping and exploration ship for hydrology work and subsequently refitted as a multi-purpose scientific laboratory with extra berths for paying passengers. We had left New Zealand with twenty members of crew and fifty-two passengers. The atmosphere on board was bathed in an easy informality, tinged with excitement.

The first few days at sea were all centred around adjustments – to the rolling ocean, the routines on board the ship and, most important, the other passengers and crew.

The ship's mess, split into two long rooms below decks at the bow of the ship with a kitchen between the two halves, was where the expedition team would meet at least three times a day to bond, swap stories of work or whatever else they had been doing during the trip so far. The first few days, a dozen or more people were laid low with varying degrees of seasickness and the mood on board was subdued.

I had not been on many ships before (and certainly not a working scientific vessel) and was, for a time, unduly impressed with the array and quality of food available to us – fresh bread and (somehow) fresh fruit, salads, several types of meat and fish every week, cakes, coffee and an endless supply of biscuits. Sailing across the Southern Ocean, the food I ate was often more varied and delicious than anything I would have eaten in a typical month at home in London. But it was richer too and I felt a little guilty eating it after a few days. At the table, my fellow expeditioners and I would sit down to meals and often gasp as the starters – steaming bowls of soup or finely dressed salads or fresh pasta with bubbling cheese – were placed in front of us by the crew.

It is perhaps an indication of how fast people get used to things – sickness, rocking boats, impressive dinners – that most of us stopped exclaiming our joy about the food after the first week. My nausea lasted for a day and a half out of Bluff, almost two days. It took a few days to get used to the continual movement but, after the initial extreme bout, it never returned. It was the same for most of the rest of the passengers – even as the sea got rougher, the number of people who were sick dropped dramatically after the first few days.

The sea got worse, as promised, a few days into the voyage. From the bridge, you could see the bow rise and fall by several storeys as it crested waves and, for several hours on several days, a pendulum on the bridge swung to indicate that the vessel was tilting by 20 degrees each side of the vertical as we made our way forward. The rocking seemed (and felt) much further from inside our cabins. From here, the sky and sea would periodically flash past the portholes as the ship tilted in each direction.

The result was a near-continuous shuddering and lurching as the ship sailed forward. Inside the ship, everything that could move, moved. Books, fruit and cups slid and rolled across tabletops as the ship tilted, coats hung from their hooks at odd angles, chairs either moved from one side to the other under decks or just fell over. My cabin was next to the ship's mess and, at least once or twice a day, I heard the metallic crash of pots, pans and cutlery pierce the bass-heavy background thrum of the ship's diesel engines. A few times, I woke up when my head banged into the wall directly behind me as I was lying asleep. I had already taped the curtain to the rail to stop it from drawing and undrawing itself all night (and rattling loudly in the process), but there was little I could do to stick myself down on my bed. If the rolling was particularly severe, I would half wake up and become aware that I was sliding up and down my narrow bunk; crashing head-first was rare but surprisingly effective at bringing me swiftly (and briefly) to full alertness, followed by a foggy, mildly throbbing head. You got used to it.

Moving around the innards of the ship, where there were no views of the outside for reference, the ship's movement manifested itself as a constantly shifting gravity field wherever you went. Walking along a straight corridor – a simple task that no one would ever think too hard about elsewhere – became the equivalent of running the gauntlet in the Southern Ocean. As the ship moved up, down and laterally around, buffeted by the enormous waves outside, the only indication inside was a subtle shift in the direction and strength of the gravity holding you down. For a while everything might be normal, the Earth's gravity pulling you straight towards the floor, as it might anywhere else on the planet. Most of the time you would also feel a slight pull to the left or right as the ship rolled through the waves. Every so often, though, the gravity would seem to double and the floor would push hard against your feet. Occasionally, and this was the oddest of all, the floor would pull away and you would feel weightless, the floor and your body falling through space at the same acceleration and neither exerting a force on the other. The last of these gravity shifts was particularly dangerous if it appeared, without notice, when you were going down stairs.

It became an unconscious act to keep hold of something whenever moving around the ship. Usually this meant the railings along the walls and bulkheads or any furniture that happened to be within reach (much of the furniture in the common areas was bolted down for obvious reasons). This didn't stop you from crashing into the walls at regular intervals or falling through open cabin doors and having to make hasty apologies to anyone inside. Taking a shower became a game of chasing a wandering stream of water around the cubicle while keeping hold of a railing on the wall. You didn't want to be washing your face with both hands if the ship lurched violently to one side and threw you against the wall or floor. This also meant that you should only apply soap to one hand at once. There's nothing more dispiriting than, in an emergency when the ship is tilting, trying to stabilise yourself by grabbing hold of a railing with soapy hands. It tends to leave you on the floor.

Moving across an open space was to be avoided if at all possible. If walking around the perimeter was not an option, open spaces required a series of confident dashes between care-fully identified waypoints where there were objects fixed to the deck. Though even that paled, in risk terms, with the logistics of carrying a hot drink from one part of the ship to another.

My legs got used to the more rhythmic movements and, within a few days, were automatically sensing the rocking of the ship and adjusting my weight to take it into account and keep me upright. When typing at the desk in my cabin, my right arm would unconsciously move out to brace me against the hull of the ship as it rocked in that direction, preventing me from tipping over, even as my chair tilted onto two legs and my computer slid across the table. After a while I no longer thought about any of this movement – of the ship or my own.

The rough seas and need for adjustment meant that even those scientists who had made the Southern Ocean crossing to Antarctica on previous occasions decided to take things easy for the first evening at sea.

By the second afternoon, on our first full day at sea, the oceanographers had started their work. Fine nets, two metres long and with a pint-sized jar at one end, billowed out behind the ship, collecting plankton or any other tiny objects that happened to be floating in their path at the surface of the ocean. Every few minutes, the nets were brought in and the specimen jars attached to the end of the nets would be taken off and stored.

Seen from the deck of the ship, the sea was a deep sapphire blue. The jars, however, came back containing a murky, pea-greenish liquid containing an uncountable number of krill and plankton, with the latter often congealed into a gooey slime suspended within the water.

Plankton bloom in the ocean from November onwards, as the Southern Ocean summer heads towards its peak and the amount of light available to photosynthetic phyto-plankton at the ocean's surface increases, allowing it to make hay. In turn, the zooplankton and larger organisms feed and grow numerous on the rich and abundant source of plant nutrients.

When plankton die, their remains fall to the bottom of the ocean, something oceanographers call 'marine snow'. In some parts of the ocean, layer upon layer of dead plankton can build up, over millions of years, to hundreds of metres in depth. In the way tree rings can tell us about the history of a forest environment, the ancient layers of plankton store information about past marine climates.

You can look at which species of plankton exist at which places in the layers, for example, or how abundant they all are and the range of specific chemicals within them. Using these clues, oceanographers can go back in time and infer details of what the organisms were eating, how populous they were and the relative abundances of various chemicals in the sea and air. All of this goes into building a model of past climate. Collecting samples of plankton behind our ship at various latitudes along our journey to Antarctica provides the crucial raw materials with which to build a baseline for these ancient climate models.

Another thing the scientists were looking for in their jars of sea water was plastic. This stuff starts as consumer and industrial rubbish thrown into the sea around cities or coastlines. Over time, because of the action of sunlight, it will start to fall apart and eventually break down into tiny particles no bigger than the millimetre-sized plankton. This artificial plankton is an indelible mark of human civilisation. Plastic like this can travel for years along ocean currents, ending up thousands of miles from its initial source. A lot of the plastic used in our goods today could end up like this, floating for thousands of years in the surface of the world's oceans. Already, scientists have identified large parts of the

North Atlantic and Pacific Oceans where particles of plastic tend to collect. In some of these places, there is more plastic than plankton in the sea surface and it is not harmless – plastic can attract algae, absorb chemicals toxic to marine life and have major impacts on the marine food chain.

Overnight, the oceanographers had also attached what looked like a small, half-metre-long missile behind the ship. This high-tech probe was filled with sensors that recorded the temperature of the sea surface, its saltiness and pH as it was dragged along behind the ship. The plan was to drag the probe behind the ship for the whole journey to Antarctica, to build up a picture of how the ocean's surface characteristics changed along the route. Every twelve hours, the probe would be reeled in, partly to make sure it had not been swallowed by inquisitive sharks or whales, but mainly to download the measurements it had taken for the previous session.

Measuring the temperature of the sea like this – in essence sticking a sophisticated thermometer into the water – is the only way to be certain what the temperature of the ocean is and how it fluctuates across a particular geographic region. Explorers led by Douglas Mawson did something similar on their journey in 1912 by pulling up buckets of sea water and measuring the temperature with thermometers. Our modern electronic probe did almost the same thing but provided thousands of data points every day. In recent decades, Earth-observation satellites have been used to measure the overall surface temperature of large sections of the ocean, from space. But it is good practice to ground-truth these measurements, to ensure the satellites are working out temperatures correctly. And anyway, no one had directly measured the properties of the parts of the Southern Ocean through which we were travelling for more than a century.

That was just the start. At one point, the oceanographers, zoologists and ecologists took turns (with the help of some of the passengers) to carry out a thirty-hour experiment as we crossed a critical boundary in the Southern Ocean. Every hour (sometimes every half hour), they threw a buoy or temperature sensor overboard. Around twenty hours in, they threw a robotic probe into the water. They were looking for the exact location of the 'convergence zone', the point at which the warm waters of the tropics give way to the cold of the poles. Here lies the Circumpolar Current, the strongest current in the world and a flow of water that isolates Antarctica from the rest of the world.

All of these incremental data points – plankton here, plastic there, temperature and salinity every few kilometres – are the bread and tedious butter of work on a scientific vessel. Many expeditions will do this for months on end around the roughest, most isolated parts of the world's oceans in a bid to understand the movement and patterns of the Earth's climate system. The Southern Ocean, in particular the track we were making along it from New Zealand to East Antarctica, is so vast, hostile and isolated that it is the least studied. But this ocean is also the most important of them all, connecting the other major ones – the Atlantic, Pacific, Indian – and playing a commensurate lynchpin role in global climate.

To track the convergence zone meant throwing twenty-eight eXpendable BathyThermograph (XBT) temperature probes into the water, one every 10–20km, as the ship made its way south. These high-tech thermometers, the size and shape of a small baseball bat, fell through the water and sent back information about temperature as they sank. They can reach 900m below the surface of the ocean before they disappear without any more of a trace.

Pairs of what looked like free-floating beach balls also went overboard every so often. These buoys started off 10m apart and were designed to float on the ocean currents for five years, sending back their position data to a satellite, by text message, every six hours.

These measurements, the first time anyone had done them, began at 56 degrees south and finished around 59 degrees south. We sailed a total of around 440km in that thirty-hour period, teams of scientists taking it in shifts to stay up and help record position and time and to keep the equipment working.

The temperature results came in fast and, once they had been plotted on a simple chart to show temperature on one axis and position on the other, we were looking at a coloured cross-section of the water across which we had just sailed. A few interesting features popped out straight away. We think of water as moving, flowing, continually mixing. Our chart was a snapshot in time but the striking thing was how defined the boundaries between different temperatures were. There was a gradual cooling of the water the further south we had gone but the convergence zone, for example, was a sharp boundary at around 56.5 degrees south and it was considerably, clearly warmer on the tropical side than it was the polar side, by several degrees Celsius. The convergence was not so much an area of gradual change as a wall in the ocean.

Also noticeable was a tongue of particularly cold water snaking in from the south. At around 100m deep it was a discrete, sharp stripe that was colder than anything around it. The only explanation, said one of the oceanographers on board, was that it had travelled all the way from the edge of Antarctica. There was no other way that water could have become so cold without being exposed to the sub-zero air temperatures of the area around the frozen continent.

This patch of water had become ultra cold in the seas around Antarctica and then travelled away, almost 800km, just under the surface of the ocean. This javelin of cold maintained its frigidity as it shot north, hardly paying attention to anything around it, passing our ship and on to who knows where else.

This was simple data and, with minimal processing, it was only a quick baseline for the experiments being carried out on board the ship. But they added another dimension to the ocean that had shaken and battered us for the past few days.

There was also a little mystery in the simple plot we saw from the temperature sensors. All of the XBT temperature probes had behaved beautifully in the water, sending back their data for almost a kilometre's depth of the upper ocean. All except one. This one had been deployed towards the end of the thirty-hour session and was working fine until it reached about 200m from the surface. At that point, the temperature of the ocean seemed to spike to 12°C for a short period. The probe continued its descent normally for another 400m and then the temperature of the water shot up to a human-body-like 37°C.

There's every chance that this was a broken instrument – it happens all the time in the field and, if true, it would be something of an achievement that there was only one of these in the entire set.

However, there was no hesitation among the scientists and passengers to come up with alternative stories for the strange data points. Could this XBT probe have encountered something in the depths? Something warm-blooded and inquisitive, which investigated the baseball-bat-sized object at 200m depth, moved away and then came in for the kill a few hundred metres deeper? A whale?

We would never know, of course, and that meant my imagination could run wild with the scant data we had. I wondered if it were something even rarer than the beautiful and unusual deep ocean creatures that I knew existed from books and television documentaries. The fantasy became more intricate the more I thought about it. Perhaps, I hypothesised, this was a creature from the imaginations and tales of the old seafarers who used to sail these parts of the world a century ago. Those people had scratched out inky lines and notes onto paper, which became the first maps of these unknown seas. But they could only account for the places they had been and, in the white spaces of which they knew little or nothing, would inscribe that most famous of legends that tells explorers there is so much more to find out: 'Here be dragons.'

THREE

On the Origin of Water

'In the beginning, darkness was hidden by darkness; all this
[world] was an unrecognisable salty ocean'

The Rigveda (1200 BC)

'In the beginning God created the heaven and the Earth . . .
And the spirit of God moved upon the face of the waters.'

Genesis, King James Bible

Before we worried about dragons, before we began
to explore the extent of the oceans and realised
there were such hidden depths in them, human
cultures had already invested water with a primordial
nature, something unknowable and beyond any possible
experience.

The Egyptians told of the time before creation, when
the sun god Atum lay on the primordial ocean, Nun. The
ancient Babylonian Gods came from the fusion of salt
(tiamat) and sweet water (apsu). For Hindus, all the creatures of the Earth came from the ocean. The Judeo-Christian
story of creation starts with the spirit of God stirring above
the waters.

In the Koran, Allah is said to have made every living

thing from water and the ancient Mayan mythology tells of 'only the sky alone is there . . . Only the sea alone is pooled under all the sky . . . Whatever there is that might be is simply not there: only the pooled water, only the calm sea, only it alone is pooled.'

Christians are baptised by holy water. Pilgrims make their way to be healed at the waters at Lourdes, Chartres and the Chalice Well. The Romans built a shrine to the goddess Sulis Minerva at the mineral springs in Bath. After they had defeated the Romans, Hannibal's armies drank from the springs in Vergèze in the south of France, which is now the source of Perrier water. Leonardo da Vinci often took to the waters at San Pellegrino in northern Italy.

In the Japanese Shinto tradition, people need to wash with water before entering a shrine. Before prayer in Islam, hands and mouths must be rinsed in water. Ancient Jews used water purified by priests before they approached an altar and individuals would purify themselves from guilt by washing their hands.

Hindus wash away their sins in the holy River Ganges, which itself comes from the Himalayas, the mountains of the gods. To this day, the cleansing power of this river manifests itself with thousands of daily bathers and culminates with the Kumbh Mela, a festival that occurs every three years in one of four cities across India. The largest of these takes place every twelve years on the confluence of the Ganges, Jumna and the (mythical) Saraswati rivers at Allahabad. The 2013 event in Allahabad attracted an estimated 80 million people over the course of fifty-five days and is one of the largest (if not the largest) gatherings of people in one place at a single time. Bathing in this festival is a sacred act, delivering people from sins and giving them direct access to heaven after death. And their connection to this vaulted celestial sphere is through water.

*

The ancients were right about one thing – water really did come from the heavens. Water is a simple molecule and the second most abundant in the universe, each molecule made from two atoms of hydrogen and one of oxygen, trios that came together, billions of year ago, in the blackness of space.

All of the hydrogen in the universe was created a few minutes after the Big Bang, in an unimaginable, fiery soup of radiation and matter. All the energy that exists in the universe we know today was, 13.7 billion years ago, condensed into a single point, with no dimensions, that exploded and began to create space and time. As it began to spread out, the universe cooled and the energy began to condense into particles and radiation. It was a maelstrom for a minute or so; there was too much energy compressed into too little space for any two particles to come too close, even if they felt some attraction. Within three minutes, however, electrons and protons had slowed enough to capture each other by mutual attraction. All of the atoms of hydrogen we have ever known – single protons circled by single electrons – were made here. And for a number of years, that's all there was. The universe expanded, cooled and hydrogen atoms bonded in pairs to make molecules that spread through the newly forming space.

These molecules did not spread through the new universe uniformly, however. Trillions upon trillions of them clumped together in countless, swirling clouds of particles that were only marginally more dense with matter than the stark vacuum around them. The growing mass of these clouds attracted more hydrogen molecules at their edges, while the molecules at their centres got more crowded. Here the hydrogen molecules smashed and bounced off each other more often than anywhere else. At some point, the rate of collisions became so high, and the pressure caused by the gravitational attraction at the centre of the growing cloud so intense, that the centre of the cloud ignited. A star was born.

In usual circumstances, two hydrogen molecules would come together through a faint gravitational attraction but, once they got within a certain distance, their outer electrons would ensure that they flew apart again. The electrostatic force between two electrons is 10^{39} times stronger than the gravitational attraction between the molecules. At the centre of a star just before it ignites, however, the combined gravitational attraction of so much mass in such a small space overcomes that electrostatic force. The molecules get closer than they would anywhere else and the protons inside them come into contact and fuse into a nucleus of the second-heaviest element, helium. That single fusion process releases a tiny amount of energy; scale this up by 10^{37} times per second for a star that is the size of our Sun – this is roughly the rate of collisions of protons in the dense centre of the star – and this output of energy makes the star shine.

For many millions or billions of years – exactly how long depends on the size of the star: the more hydrogen it started with the faster it will burn – the molecules at the centre will continue to fuse into helium. The star is finely balanced now – the gravitational force of so many atoms wants to pull everything closer but the energy released during fusion counteracts that force to prevent the star from imploding. Once all of the hydrogen has fused into helium in the star's centre, though, it can no longer resist the immense gravitational pressure and the core begins to collapse. This heats up a shell of gas around it, so much so that hydrogen there begins fusing. The outer shell of the star expands to several times its original diameter, cools down and glows red. The star has now become a red giant.

Meanwhile, at its core, the star shrinks until its helium molecules (left from the original hydrogen-burning phase of the star's life) are forced close enough together to fuse into atoms of oxygen and carbon. The star's outer layers get hot again, glowing blue and white.

The biggest stars go through several stages of fusion, and create all the heavy elements we know of, everything that is more complex than a simple hydrogen atom – starting with helium, and progressing all the way up to iron, in a process known as nucleosynthesis. Once a big star's core is made from large amounts of iron, it can no longer continue the process of fusion and, at that stage, the star is as good as dead. Fusion stops, and the centre of the star collapses into a white dwarf, a super-dense object that comprises half the mass of the original star, but is no larger than the Earth. A teaspoon of this object would weigh a ton on our planet. If the star is big enough, the centre will shrink further, to a point, a singularity from which there is no escape, and form a black hole.

The outer layers of the star's atmosphere, meanwhile, are blown away in an explosion that is, momentarily, brighter than the rest of the galaxy. Around the white dwarf at the centre will be a halo of gas and dust containing the raw materials for the formation of planets in future, including carbon, oxygen, neon, sulphur, sodium, argon and chlorine. The remnants of low-mass stars, called planetary nebulae, are among the most beautiful objects in space. The hot core of the star lights up the surrounding gas clouds, producing vivid fluorescent colours. Scientists have been moved to give these nebulae names such as the Cat's Eye, Starfish Twins, Blue Snowball, Eskimo and the Ant.

The first time any of this happened, the universe was already many millions of years old, and contained all the ingredients needed to make water. All that was left were the right conditions to bring these ingredients – hydrogen and oxygen – together in the right way and let chemistry take its course.

To do that the universe needed to make new stars from

the ashes of the old. Unlike the first generation of stars that were made in the years after the Big Bang, from virgin clouds of hydrogen floating through the universe, our Sun made its first baby cries while engulfed in the clouds of dust filled with a rich array of heavy elements, the remnants of a first-generation star that had gone through all the fusion it could muster and then exploded. As our Sun got hotter and spewed out radiation, it energised the hydrogen and oxygen atoms that happened to be floating in the vast elemental clouds around it. These two elements found themselves slamming into each other and, instantly, they combined – and a molecule of water was born.

That process happened in the early years of almost all of the second generation of stars in our galaxy. The energy thrown out by a new star helped the hydrogen and oxygen already in its vicinity to combine. These second-generation (and beyond) stars act as giant hoses, pumping unthinkable volumes of water out into the blackness of space at hundreds of thousands of miles per hour. The water emerges as jets of superheated gas that, once they get far enough away from the heat of the young star, condense into an uncountable number of water droplets and, eventually, ice.

Using the Herschel space observatory, scientists have seen these stellar firehoses in action in the northern constellation Perseus where, every second, a protostar is pumping out more than a hundred million times the volume of water that flows through the Amazon, watering the void of interstellar space at an astonishing 200,000kph.

Vast clouds of water crop up everywhere in our galaxy, the by-product of all sorts of stellar processes, intermingling with dust and other gases, being torn apart and recreated and sometimes coming together to freeze into trillions of giant comets. (The biggest cloud of water vapour ever discovered by NASA scientists was amassed around a black hole that is 12 billion light years from Earth, 20 billion times more massive than our Sun. Around it sits a water reservoir equivalent to 140 trillion times all that in the Earth's oceans and spanning hundreds of light years across. The water sits at -53°C and the cloud is 300 trillion times less dense than the Earth's atmosphere.)

But these water bullets around the protostars have a lonely, dismal fate ahead of them. They are sent off into space, to wander far from their source until, eventually, the bonds between the hydrogen and oxygen atoms get broken apart by the intense, energetic ultraviolet radiation that criss-crosses deep space. Fortunately the water we know is formed in a different way.

The water that will end up on Earth, at this moment in the evolution of the solar system, has yet to form. Even as the water bullets were being fired out of the poles of our Sun, something more long-term was happening in the clouds of matter around our baby star. We know that these clouds were more varied, dirtier and more interesting than the clumps of plain hydrogen that populated the early universe. Beyond a certain distance from the surface of the newly forming star, the gravitational attraction of the Sun kept the dust and gas in orbit but did not pull the material into the star itself.

In this nascent solar system, beyond the Sun, individual molecules and atoms floated between vastly bigger dust grains (big in atomic terms but still only a millionth the width of a human hair) made from carbon, silicon and other elements. In the coldest parts of the cloud, the temperatures were barely above absolute zero and the particles and atoms moved sluggishly, a few fractions of kilometres per second. And there wasn't much stuff around – the density was around 10,000 particles per cubic centimetre. Compare that with an ultra-high vacuum in a laboratory, which has of the order of 10^{10} particles per cubic centimetre, perhaps 10^8 if you have a really, really ultra-high vacuum. What an astronomer calls a dense cloud is still a million times less dense than what we call an ultra-high vacuum on Earth. It is really very empty in space.

Most of the material in that empty space was gas, and that gas was mostly hydrogen. One in 2,000 particles was an oxygen atom and one in 10^{12} was a dust particle made from sand-like silicates or graphite-like grains, remnants of dying stars.

On average, one hydrogen atom landed on a dust grain about once per day. Given their tiny size and mass, hydrogen atoms would often evaporate away from the grains almost as soon as they had landed. Oxygen atoms tended to stick around for a bit longer. Randomly and very rarely, atoms of both oxygen and hydrogen would strike these grains of dust and, even more rarely, they would do so at the same time and for long enough and be close enough on the dust grain to form chemical bonds with each other.

Each water molecule we know of on Earth started its existence on one of these dust grains, when an oxygen atom and two hydrogen atoms happened to land themselves on the dust and began to share their outer electrons on their new home. Over the course of hundreds of thousands of years, as it tumbled through space and collided with more hydrogen and oxygen, the grain of dust acquired successive layers of molecules that turned into ice, until it doubled in size thanks to a mantle of ice (though this was alien ice, nothing like the crystalline, translucent material with which we are familiar).

These dust grains floated in clouds of their own for many millions of years. But much like the clouds of hydrogen molecules that were brought together by gravity and turned into stars at the earliest stages of the universe, the ice-encased dust grains wandered through space in ever-denser clouds. The grains got closer and the increasing gravity pulled in other dust grains and atoms circling at the edges of the cloud. Forming these clouds was a crucial step in preserving the water: the intense UV radiation in open space would have torn apart the water molecules on the surfaces of many of the grains or prevented the oxygen and hydrogen atoms from sticking onto the grains in the first place. By gravitating into clouds, the ice-encrusted grains in the deeper regions were shielded from the UV by the grains on the outer edges, which absorbed and scattered the radiation before it could reach the inner regions.

The destruction wrought by the UV on the grains was, by the way, how we first spotted these vast clouds of cold water vapour that swaddle baby stars. TW Hydrae is an orange dwarf star in the Hydra constellation, 10 million years old, somewhat smaller and cooler than our Sun and 175 light years from Earth. Surrounding it is a swirling disk of dust nearly 200 times the distance between the Sun and the Earth, all of it bathed in a haze of water molecules that have formed on dust grains and been shorn off by the UV radiation. The light signature from this thin cloud of vapour was detected by the scientists at the Jet Propulsion Laboratory in Pasadena, using the Herschel space observatory, for the first time in 2011. This star, nothing out of the ordinary and which will develop into a solar system in the next few million years, shows that the conditions to create water-covered planets like ours are probably routine in our galaxy.

Several billions of years ago, in a similar dust cloud surrounding our forming Sun, the ice-encrusted dust grains that survived the UV rampage began to get closer in the relatively dense centre of the cloud, sticking together to make slightly bigger grains. This was a delicate process and not all the collisions led to the dust grains combining – whenever they collided with too much velocity, the grains would shatter and it would be back to square one for the water molecules, which would fly off uselessly into space. When the dust grains met less violently, the particles would stick. The ice often helped here, acting like a glue. The individual dust particles grew, first to a few millimetres across to form tiny stones, which then combined successively into rocks, boulders, asteroids and, eventually, planetismals that were several kilometres across.

This is an important moment because the kilometre-sized planetismals began to interact with each other through gravitation. Before that time, all the dynamics had been determined by the star, the most massive body in the system. At the kilometre size, the rocks started to feel each other's gravity, leading to more collisions and amalgamation until they made the beginnings of a protoplanet – something typically the size of the Moon, a few hundredths of the mass of the Earth. These objects, with all their minerals and ice, were the building blocks of our planet.

All of our solar system's planets and the near-infinite balls of rocks and ice formed, in this way, phoenix-like, from the random dance of ashes from a star that died and exploded more than 5 billion years ago.

Before there was a solar system, the water that is now on our planet and in our bodies floated in the void of space. It took around 20 million years for the proto-Earth to coalesce from the clouds of dust and ice spinning around the young Sun. This early Earth, 4.5 billion years ago, was a ferociously hot place. The surface was covered in volcanoes, much of the ground ran with molten magma, and huge rocks struck the surface on a regular basis. One of these collisions came in the form of an asteroid the size of a small planet, which gouged out a chunk of the Earth's crust and mantle that began orbiting our planet as the Moon. Underground on the Earth, the decay of radioactive elements produced enormous heat. There is a reason why these first half billion years are known as the Hadean era, named for Hades, the hellish underworld of the ancient Greeks.

Most, if not all, the water on the surface of the Earth at this time came from the rocks and ice that had coalesced to form it in the first place. But the early planet had trouble keeping hold of these water molecules. Without a fully developed atmosphere, they would have escaped the Earth and boiled off into space.

Once the planet's mass was largely in place, the atmosphere stabilised. All the while, water was being pushed to the surface by the colossal geological processes that gave Earth its internal structure. Heavy elements like iron largely flowed to the centre, and the distinct layers of crust, mantle and core we see today began to form. Water and other volatile compounds from the rocks were driven upwards as the mantle cooled. Volcanoes and other fissures in the crust allowed superheated water vapour to escape into the atmosphere.

Despite the high temperatures on the Earth's surface – more than 200°C – pools of liquid water are thought to have existed there in those first 100 million years, thanks to the immense pressure of the atmosphere, which was rich in nitrogen, water vapour and carbon dioxide. A few hundred million years after the birth of our planet, the atmospheric pressure had dropped (thanks to falling carbon dioxide levels), and the temperature had dropped too (for the same reason). Now the Earth was cool enough for liquid water to stay put. At this point, somewhere around 4 billion years ago, the water vapour in the air began to condense out and it rained. And rained. Possibly for millennia. If nothing else, the deluge recounted by numerous mythical creation stories correlates with what happened in the earliest, most tumultuous years of the Earth.

Proving every fine detail of this history is difficult, as you might expect for something that happened under such hellish conditions, so long ago. Much of the evidence written into the rocks created at the time was erased as they were recycled through the Earth's crust and mantle over the eons since. We do have a restricted window into this time, however, in the form of tiny crystals known as zircons, which formed 4.4 billion years ago and survive to this day. These zircons contain the correct types of oxygen in the right ratios to show that liquid water was around when they formed. (In contrast, the oldest sedimentary rocks, which would have needed water to form and are a more abundant evidence of the liquid, are only 3.9 billion years old). This theory of the origin of Earth's water is called the 'wet' formation theory.

There is, however, a conundrum that still plagues planetary scientists about the early years of the Earth. Our planet formed inside the so-called 'snow line' of the solar system, a region with a radius from the Sun that is 2.7 times the star's distance from the Earth. Here, any ice would have quickly sublimated into gas. That meant that the planetesimals that came together to form the Earth would have become relatively dry, certainly far drier than the amount of water we see today would suggest. Where did our Earth's extra water come from?

The consensus (though by no means everyone agrees with it) is called the 'dry' formation theory and the argument goes that the extra water came in from the outer solar system; in other words, the Earth started from dry planetesimals and the water came later. Around 4 billion years ago, at the same time or just after the deluge was happening on the early planet, the inner planets of our solar system were seemingly pummelled with comets and asteroids and the evidence for these events, known collectively as the Late Heavy Bombardment, are carved into the surface of the Moon. No one knows how many objects hit the Earth and how much water they brought but this period of intense bombardment is thought to have lasted from 4.5 billion to 3.8 billion years ago.

The water brought to Earth at this time would have had to come in quickly, shielded inside large objects such as comets or water-rich asteroids. Any water at the surface of the objects would have evaporated as they approached the early Earth but, a few metres under the surface, the precious cargo would have been preserved.

The dry theory has several unknowns. For example, where did the objects that collided with the Earth come from? Perhaps one of the gas giants such as Neptune wandered into the path of a ring of comets in the outer solar system and caused huge chunks of ice and dust to fly in all directions? If only a quarter of the bodies that hit the Earth were comets – made up of half dust and half water ice and often described disparagingly as 'dirty cosmic snowballs' – that could account for all of the water in our planet's oceans. It would also explain why the heavy bombardment period lasted so long – anything originating so far out in the solar system would have taken a long time to orbit and reach the inner planets, so the impacts would be stretched out over hundreds of millions of years.

A wrinkle in the comet theory is that evidence from recently studied comets – Halley, Hyakutake and Hale-Bopp – shows that these objects have a high proportion of an isotope of hydrogen called deuterium. Though deuterium largely behaves in the same way, chemically, as hydrogen, it is heavier since it has a neutron as well as a proton in its nucleus. Compared with the water in the Earth's oceans, these comets had double the amount of deuterium compared with hydrogen. Put simply, this extraterrestrial 'water' doesn't match what we see on Earth.

The latest blow to the comet theory came with some of the first results of the European Space Agency's Rosetta probe in late 2014. Rosetta spent 10 years hurtling through space to catch up to the comet 67P/Churyumov-Gerasimenko, one of the Jupiter family comets, which is currently around

300 million miles from Earth. Using an on-board spectrometer called Rosina, Rosetta's scientists could analyse the water flying off the comet and measure the ratio of heavy water to normal water. Rosina found around three times more heavy water compared to normal water on the comet than there is on the Earth. Because the values do not match, that means the water on our planet was not likely to have come from comets such as 67P/Churyumov-Gerasimenko.

If these comets are representative of the early solar system (and there is little reason to think otherwise), then we need to keep looking to find the source of the Earth's water.

Another idea is that the water came in the form of asteroids originating in the region between Jupiter and Mars, between 2 and 3.5AU (1AU – or astronomical unit – is the distance from the Earth to the Sun). The outermost part of this asteroid belt contains water-rich rocks known as chondrites and the theory is that the Earth collided with one or several planetismals from this region. Additional water could have come from asteroids on highly tilted orbits that arrived during the heavy bombardment period. This source of water is also a potential solution for the deuterium problem since it proposes that no more than 10 per cent of the Earth's water came from comets that started life in the outer solar system. This combination of water from the nearby asteroids and faraway comets seems consistent with the deuterium-hydrogen ratio we see on Earth.

Wherever the water came from, whether it came in the violence of bombardment (the dry theory) or the more leisurely accretion (the wet theory), the payloads of water turned into our nascent oceans. The next challenge faced by our young planet was to hang on to that water.

The Earth was helped in this regard by its positioning from the Sun in the habitable zone, where it was neither too hot nor too cold for liquid water to exist without trouble.

Imagine the inner solar system just after the Late Heavy Bombardment – the Earth would not have been the only planet to be pummelled by water-bearing rocks and snowballs; Mars and Venus would also have been loaded up with water during this time. Both of those planets are completely devoid of running streams or oceans today. What happened?

On Venus, the intense solar radiation would have created a humid world where water vapour would have reached all the way to the highest reaches of its thick atmosphere. The higher the water went, the more likely it was to encounter energetic ultraviolet radiation coming from the Sun, whereupon the molecule would have been torn apart. The hydrogen, being so light, would easily have then escaped into space and a vital ingredient in the formation of water would have disappeared from the planet. Fast-forward billions of years and we are left with a planet devoid of this life-giving molecule.

Water molecules too far from the Sun, conversely, would end up freezing. When there isn't enough solar energy to keep rivers and oceans moving, a planet can enter a state of runaway glaciation. Polar ice caps will expand and, because frozen water is white, they will increasingly reflect away any sunlight that does land on the surface. This would, in a vicious cycle, cause the planet to get even colder. This is probably what ended up happening on Mars, which is just outside the liquid-water zone from the Sun today. There is evidence that water did flow on the surface of the red planet at some point in its history – pictures from NASA's Mariner, Viking and Global Surveyor probes show what look like channels on the oldest portions of the planet's surface and large flat plains in the northern reaches that resemble the Earth's sea floor – but Mars left the habitable zone at some point in its history and the running water disappeared.

The Earth also had to overcome another problem in keeping the water on its surface liquid in its early years. By the time the Sun stabilised, 3.8 billion years ago, it was giving out around a third less energy than it does today. Never mind technically being in the habitable zone, the reduced amount of energy falling on its surface, compared to today, would have caused the surface water to freeze. Geological records, though, show that the water was not frozen and, more important, that life had already appeared.

This 'young Sun paradox' can be explained by understanding the changing atmospheric composition of the Earth over the past few billion years. The atmosphere of the early Earth contained much more carbon dioxide and methane than there is at present. These greenhouse gases trapped energy on the planet's surface, keeping the Earth warmer than it might otherwise have been.

Even if the glaciers had overcome the surface of the Earth (and this indeed has happened in the planet's history in the form of several ice ages), our planet has an in-built mechanism to bring the flowing water back.

In a natural cycle called carbonate-silicate weathering, rainwater absorbs carbon dioxide from the atmosphere to create carbonic acid that can disintegrate silicate rocks over geological timescales, making soil. These weathered rocks and soil wash out into the sea and the carbon dioxide sinks to the bottom of the ocean in the form of calcium carbonate. Over millions of years, the shifting continental plates push the calcium carbonate into the upper mantle, where it undergoes chemical reactions and is eventually re-released into the atmosphere via volcanoes.

If the Earth ever got to a stage where glaciers covered the surface and the oceans were frozen, the carbonate-silicate cycle would be halted because of a lack not only of rain but any ocean into which the rocks could erode. The Earth's tectonic plates would continue to move, however, and volcanoes would still continue pumping out carbon dioxide. Over millions of years, the levels of this gas would rise inexorably and the greenhouse effect would be large enough to melt the glaciers. In this hothouse, the warm oceans would release lots of water vapour into the atmosphere, causing colossal rainstorms that would speed up the erosion of rocks, thereby pulling carbon dioxide back out of the atmosphere again and into the bottom of the ocean.

With all the pieces in place – positioning in the solar system, asteroid and comet bombardment, plate tectonics, volcanoes – water arrived on the young Earth and our planet was formed in just the right way to keep the vast liquid oceans that we see today.

FOUR

The Hydrosphere

We have not known our oceans for long. Since the fifteenth century, ships have criss-crossed the Earth, plying trade routes and exploring for new wealth and lands. Merchants, armies and others sailed local seas and skirted coastlines for millennia before that. But, throughout all that naval history, the seas were no more than a background, an obstacle that was there just to be crossed in search of more interesting lands beyond.

Sailors and fishermen knew there was life in the upper reaches of the water but the prevailing thought, until as late as the mid-nineteenth century, was that nothing could exist much deeper than a few fathoms. The cold, dark, high-pressure world of the ocean depths was probably a lifeless place, so there was little reason to pay much attention to it. Even as the oceans became the highways for the new age of exploration, no one seemed to wonder how much water there was and how deep it all went, or concerned themselves with how complex or structured that world might be.

That shroud over the ocean began to lift in the early 1870s, after two British naturalists proposed an expedition to systematically navigate and survey the world's oceans.

William Benjamin Carpenter and Charles Wyville Thomson had already been on short cruises to explore the deep ocean around the coasts of Scotland and the Faroe Islands in the late 1860s, confirming previous hints from scientific colleagues in northern Europe that organisms did exist much deeper than anyone had ever thought, more than a kilometre below the surface. Those initial discoveries ignited Carpenter and Thomson's interest in looking for life in the deep waters and they wanted to know how far it went. Together they proposed to the Royal Society of London and the UK government that they fund an expedition to chart the oceans to a level of detail never seen before. Their proposal came at a useful time for communications companies that wanted to lay telegraph cables from Europe to the United States (something that required a good knowledge of the shape and depth of the sea floor) and so the two naturalists got the £200,000 they needed – more than £10 million in today's money – to launch the Challenger Expedition.

The expedition, a four-year journey around the world to chart, survey and measure the oceans, was named after the HMS *Challenger*, a 200ft-long, square-rigged former warship lent to the expedition leaders by the Royal Navy. The wooden ship had had most of its cannons and ammunition removed and the decks had been refitted with laboratories and storage space. Captain George Nares took charge of more than 200 crew members while Thomson supervised the six scientists on board. The ship launched from Portsmouth in December 1872 and, starting at the Canary Islands in February 1873, the scientists took the same set of measurements and samples at 362 waypoints on their journey. By the time the HMS *Challenger* got back to England in 1876, she had travelled almost 70,000 nautical miles across the seas around North and South America, South Africa, Australia, Japan and islands in both the Atlantic and Pacific Oceans.

The main spur for the expedition had been to collect and

catalogue marine life – Thomson's scientists dredged the sea floor and collected samples from the water using nets. Everything they caught was carefully sorted, documented and preserved in jars filled with alcohol. The expedition's on-board artist, Jean Jacques Wild, made intricate sketches of many of the almost 5,000 new species that his shipmates discovered in the depths – including new sea cucumbers, rays, jellyfish and floor-dwelling foraminifera – as well as the places and people the ship visited along its way.

But, crucially for the study of the ocean itself, the *Challenger* scientists also recorded the physical properties of the water around them, everywhere the ship stopped. They lowered thermometers into the water to record the temperature of the ocean at varying depths, they threw weighted logs overboard to track the direction and speed of the currents and they brought samples of sea water on board for chemical analysis. The *Challenger* also had almost 150 miles of rope on board for sounding, a method of working out the depth of the ocean at a particular point that involved attaching a weight to the rope and lowering it overboard until it hit the sea bed. The rope was marked by flags every few fathoms and counting the number of flags that went into the water gave the scientists a good measure of the depth at a particular point. A small cup attached to the bottom of the rope would also scrape the ground and collect a sample of the mud for later analysis by the scientists on board the ship. The deepest measurements on the expedition were taken at more than 8km below the surface, in the south-west Pacific Ocean between Guam and Palau. We now know that that spot is close to the deepest point ever measured in the oceans, 11km from the surface of the water and near the southern tip of the Mariana Trench – a spot named Challenger Deep after Thomson's expedition.

On their return to England, Thomson wrote that the *Challenger* had come back with '563 cases, containing 2,270 large glass jars with specimens in spirit of wine, 1,749 smaller stoppered bottles, 1,860 glass tubes, and 176 tin cases, all with specimens in spirit; 180 tin cases with dried specimens; and 22 casks with specimens in brine'. That material took more than twenty years to compile fully and the resulting analysis was published in a fifty-volume series, totalling almost 30,000 pages. That biological data provided material for scientists to study for decades afterwards. But, as we will see, it is the physical measurements of the ocean that Thomson's team made which are still reverberating through modern climate and oceanographic research to this day.

The Challenger Expedition kicked off the new science of oceanography. The blank, unknown space between continents quickly became a topic of interest in its own right and, in the twentieth century, ships from around the world began to explore and examine the reach and the depth of the oceans.

The ship taking us to Antarctica, the *Akademik Shokalskiy*, was named after one of the intellectual heirs to the scientists aboard the Challenger Expedition, the great Russian oceanographer Juliy Mikhailovich Shokalskiy. He had enrolled in the Naval College in St Petersburg a year after the HMS *Challenger* had set sail and, despite suffering from seasickness during his student voyages, began service with the Baltic fleet of the Imperial Russian Navy four years later. He studied the seas he sailed and conducted work on everything from mapping to hydrology, meteorology to glaciers. He rose through the rank of the military to lieutenant-general and his scientific work led him to become president of the Russian Geographical Society by 1914. Three years after that, he published his most important scientific work, 'Oceanography', where he introduced the idea that the world's oceans were all interdependent, a 'world ocean' as he called it. His work made him the *de facto* father of oceanographic sciences in Russia and, after dozens of expeditions and hundreds of research papers, he lived out his final years as a teacher at Leningrad University, until his death in 1940.

Expeditions led by Shokalskiy and others in the early part of the twentieth century found more life in the depths than anyone ever expected, new species and new habitats, many living in surprising underwater environments that would previously have seemed too extreme, impossible for life. Most important, however, the measurements of water temperatures and currents by Shokalskiy, Thomson and other physical oceanographers kick-started our understanding of how the water on the surface of our world behaved.

The movement of water in and around the oceans, and the ways in which the ocean interacts with the atmosphere, is at the heart of the Earth's climate and weather. The oceans absorb and distribute the Sun's energy, dispersing nutrients and gases in a way that make the world habitable. The hydrosphere is a profound, symbiotic link between the physical body of the planet and all of the life on it. In fact, for the purposes of life, the oceans are the Earth.

Before we look at that in more detail, let's first take stock of the water on our world. In absolute terms, there are around 1.5 billion cubic kilometres of water on the Earth, making it the most abundant single substance in the biosphere, covering around 71 per cent of the planet's surface to an average depth of 3,700m (though, as we have already seen, the deepest point at the bottom of the Marianas Trench in the Pacific Ocean is much deeper than that). Most of the water, around 97 per cent, is salty and the vast majority of rest (another 2.1 per cent) is locked up in the polar ice caps and glaciers. Almost all the remaining fresh water sits in the ground, either in the soil, in lakes or in underground aquifers. A vanishingly small amount (less than 0.001 per cent) exists as water vapour in the atmosphere.

If all of the world's ice and snow were melted and spread evenly over the surface of the Earth (around 510 million square kilometres), the depth of the resulting ocean would be between 50–120m. Using the same technique can help us compare the relative amounts of other water on Earth: fresh groundwater would create a global ocean around 15–45m deep, while the water from the world's lakes would only rise to around half a metre. The water in the atmosphere, in the form of vapour and clouds, would cover the Earth to a paltry 0.03m. This small figure belies its importance, however, since that tiny amount of water in the atmosphere is the driver and carrier of the world's weather.

Despite the sheer quantities of water on this planet, only 1 per cent of it is available for human (and other animal) use. This is all we have to wash, drink, water our crops and use in our factories. And we share that sliver of water with every other living thing here.

You probably drew the water cycle at school. The Sun warms the sea and plants release water from their leaves. Water evaporates and the vapour rises into the atmosphere where it cools, condenses into clouds and is blown around to different parts of the Earth. Clouds contain tiny droplets of water or crystals of ice that remain suspended in the atmosphere as long as they are small enough and air currents can keep them afloat. Over time, the droplets and crystals gain more water and, eventually, gravity wins, causing them to precipitate. If the clouds are over cold places, the water will fall as delicate snowflakes, collecting on the polar ice caps or building glaciers.

But mostly the clouds will shed rain. A tiny fraction of this fresh water might be used by plants to grow, or by animals (including humans) to drink. Most of the water that ends up on land will either run back into the sea, collect in lakes, or penetrate the ground to feed aquifers. On the path to the sea, rivers distribute important nutrients and minerals around the biosphere and create fertile new lands for animals, plants and agriculture. Over time, water's corrosiveness has also shaped landscapes by dissolving rocks, carving great channels and creating vibrant habitats.

The hydrological cycle, to use its technical name, is the largest movement of any substance on Earth. It might seem that something as obviously simple and so important for human life and society might have been a kind of folk knowledge, built up over several thousand years through casual observations and then codified and improved as instruments and scientific thinking got better. It might surprise you, therefore, that for most of recorded history we did not understand the water cycle at all.

Civilisations in Sumeria and Egypt that existed more than 5,000 years ago relied on the rise and fall of water to keep their societies running. They built irrigation mechanisms to manage and exploit their resources and created social and political hierarchies based around the access and control of that water. But there is little evidence that the engineers of the time were aware of the global cycle of water and how it appeared in their rivers. There are hints from Chinese settlements some 3,000 years ago that they had inklings of how different bodies of water might be related – they noticed that their wells rose and fell in relation to the tides, for example, and some of them referred to rivers as 'threads' of the sea.

But exactly who came up with the idea that the Earth's oceans and rivers were connected via clouds is unclear. The concept of some kind of cyclical behaviour to the world's water seems to have emerged around 3,000 years ago in the work of the early Greek philosophers. Thales of Miletus, who lived around 600BC and was one of the earliest thinkers who we might now consider to be 'scientific' in his attempt to get to the bottom of everyday life, believed that water was the origin of all things and that everything would eventually return to water. One of his followers,

Anaximander, described the process of evaporation as the way water moved from sea to sky. Others, including Plato and Aristotle, subsequently added ideas that clouds could move water through the air and then produce rain to feed rivers and streams, and they provided descriptions of how water might also then seep into the ground. Most of the elements of the water cycle seemed to have been in place but, unfortunately, the Greeks ended up connecting the dots the wrong way around.

Thales, for example, correctly proposed that the water in rivers was connected to the sea. But he also thought winds forced sea water into the Earth and then somehow upwards into mountains, from where rivers flowed back into the sea. Others suggested that the Sun lifted water from the sea and then that water rained into underground reservoirs on the land, which in turn fed rivers. Some reservoirs were bigger than others, they thought, explaining why some rivers were perennial and others were not. Plato took this idea to an extreme, proposing the existence of a vast network of underground reservoirs and channels, which he called Tartarus, from where all the waters of rivers and seas came.

This flawed understanding of the water cycle was no big problem for the rapidly growing civilisation of the time and it did not prevent cities and empires from expanding. Around 2,000 years ago the Roman Empire was filled with engineers who designed aqueducts and other complex ways to exploit their available water resources. We will come back to that utilisation of water by early societies in chapter six, but the attempts to further understand the global water cycle stagnated for more than a millennium, until it was picked up again during the Renaissance years in Europe.

Leonardo da Vinci studied water and its movement obsessively and laid down many of the founding scientific ideas around hydrology in his writings. He had proposed, by the start of the sixteenth century, that the water in rivers came from rainfall, and also that water moved through the Earth in a similar way to how blood circulates in animals. Leonardo was influential at the time and his ideas on water took hold in the writings and thought of others. Slowly they began to dismantle the incorrect ideas of Thales and Plato. Out went underground reservoirs, Tartarus and any thoughts that water might seep from the sea, through the Earth and up mountains. A century of measurements by scientists including Pierre Perrault, Edme Mariotte and the famed astronomer Edmund Halley (who became interested in the water in the air when he noticed his telescopes, positioned on a mountain on the island of St Helena, getting fogged up with condensation) confirmed that water condensed from air as rainfall and caused streams to flow. By 1800, the English physicist and chemist John Dalton had measured the annual rainfall across England and Wales and shown it was enough not only to fill streams and rivers, but to seep into the ground too and therefore account for the evaporation from there.

We now know that the hydrological cycle transfers water from one part of the world to another, purifies it for use by plants and animals, replenishes the land with fresh water and transports minerals and nutrients around the globe. Leonardo da Vinci was not far off in calling it the Earth's blood system.

Different molecules of water will take different amounts of time to go through the full cycle and will make their way back to the sea in different ways. A droplet on the surface of the ocean might evaporate (leaving behind its salt), end up condensing into a cloud, and raining down back into the sea in a matter of days. Another droplet could find itself in a cloud that gets blown far inland, is then waylaid on a mountainside for a few weeks, before falling into a river and running back into the sea over the course of further weeks. If the droplet falls far from a river or lake it could seep deeper into the ground, though porous soils and rock until it hits an impermeable layer of the earth. There it might collect in an aquifer, where it could sit for any amount of time up to hundreds or thousands of years. Perhaps it will return to the cycle when it is tapped by humans for their drinking or agricultural needs or sucked up by plants and trees. Some aquifers also inch their way towards lakes or oceans at a rate of anything from a few centimetres per day or per year. As for the clouds that form and end up precipitating snow over cold regions such as the poles, they might lock up water molecules into ice sheets and glaciers that stay frozen for hundreds of thousands, perhaps millions, of years.

The oldest known water body in the United States, the Ogallala Aquifer, contains around 3.6 million cubic kilometres of water trapped in rocks underneath eight of the Great Plains states. Geologists think it formed alongside the Rocky Mountains, which would mean the water has stayed in place for between 2 and 6 million years. Aquifers such as this are sometimes said to contain 'fossil' water, an indication of how old the water is rather than because anything is actually getting fossilised underground here.

Over the course of 3,000 years, on average, the Earth's hydrological cycle, moving water from the surface to atmosphere and back, processes an amount of water equivalent to all of the world's oceans. We know this movement of water as weather.

The first time I remember feeling the monsoon rain was during a visit to my grandparents' home in a small town in north-east India, near the border of Nepal and around a hundred miles south of the Himalayas. I was fifteen years old and, unused to the heat, had spent an uncomfortable few weeks in July doing my best to avoid the muggy air and burning sunshine that followed us everywhere we went while visiting far-flung members of family. Seeking respite, I'd sit and read on a wooden bench that was jammed up against the walls of a wide verandah that my grandfather had built across the front of his bungalow. A main road lay a few hundred metres in front of the house and beyond that and halfway to the horizon there was a railroad. I'd often stop to watch the scarlet trains clack by on the tracks, most of them covered in people clinging to the outside like barnacles on a boat.

One sleepy afternoon, I noticed it was getting dark a little earlier than I had expected. I looked out to see bulbous, dark clouds above the railroad. In fact, the clouds filled the previously blue sky in all directions. A warm wind blew across the verandah, strong enough to snap several of the shutters shut at the front of the house. The sail-like leaves on the coconut trees at the bottom of the garden hissed and swayed in the air. The smell of earth filled my nostrils and the air was thick, pregnant with possibility.

The world seemed to pause for a moment before the first drop of water appeared. It didn't seem to fall from the sky but, instead, materialised from the humid air itself, a fat splash of water that met its end on the stone path in front of the house. Another drop soaked into the dry earth nearby a moment later. The leaves on the coconut tree began to sound a syncopated crackle just before, all at once, the skies became liquid.

Lightning flashed repeatedly nearby, followed by loud cracks in the air that sounded like splintering trees. Frogs, lizards and worms emerged from their usual hiding places in the undergrowth and took shelter wherever they could under the eaves of my grandfather's house. The rain cascaded off the roof of the verandah, a never-ending curtain of water that joined a gushing river that had started flowing along the garden path just moments before. Outside on the road, those caught unawares ran along the tarmac towards the nearest shelter, newspapers or plastic bags covering their heads, a relatively pointless attempt to stay dry in the face of the deluge.

Others, curiously, were running into the rain. They smiled, shrieked, laughed and sang as they got drenched. For some reason I felt compelled to do the same. I kicked off my shoes and stepped out from under the safety of the verandah. The warm rain soaked my arms, my T-shirt, my shorts and my legs. I spun around as the refreshing rain washed away the fug of summer. All around was noise – people shouting, frogs croaking, birds singing. The energy of this water had woken up the whole world.

The monsoon is a life-giving part of the water cycle that moves across India from June to September. It is immortalised in poetry, epics and novels for good reason – the vast majority of the water that falls on India does so between these months and the monumental amounts make the sub-continent bloom with life. Tracts of land that turn dusty and arid over the preceding winter become verdant oases soon after the rains arrive. Most Indian farms are rain-fed and the monsoon has an important role in flooding the all-important rice fields across the country and watering the cotton, grain and many other agricultural plants and animals that provide much of the country's food and the basis for its economic output.

The rains occur because of an imbalance in the temperatures of the land and ocean around the Indian sub-continent. As the Sun's energy hits the surface of the Earth, it heats both the sea and the land but, because water has a much higher specific heat capacity than land (in other words it takes a lot more energy to raise its temperature by a single degree), the land ends up much hotter than the sea after a few months. In particular, the Thar Desert and its surrounds in northern and central India heat up considerably and the air above those territories expands, creating a low-pressure area in the atmosphere. Above the nearby Indian Ocean, meanwhile, the atmosphere stays at a higher pressure because the water has not become as hot as the land. The atmosphere cannot sustain an imbalance like this for long. In an attempt to level the scales, the monsoon wind blows north from the moist, high-pressure atmosphere above the water and into the dry, low-pressure air above the desert. The wind reaches as far as the Himalayas, where it cannot move any further north and, instead, it rises high into the atmosphere. There it begins to cool and the copious water in the moist air starts to precipitate out as a deluge of rain.

In a microcosm, that is how all of the weather systems in the world operate. What we call weather is largely a result of water vapour moving around temperature and pressure imbalances in the troposphere, the 20km-thick lowest layer of the Earth's atmosphere. Almost all of the water vapour in the atmosphere is contained here and, rather than in pillowy or streaking clouds, most of it moves invisibly through the air. Depending on the place, temperature and winds, there might be up to fifty times more water vapour in the air than in clouds.

That movement of water vapour is the Earth's way of redistributing the Sun's energy as it shines unevenly across our planet. Sunlight comes in directly from above to someone standing at midday at the equinoxes at the equator and it is at its most intense there. However, it comes in at a shallower angle the further away you are towards the poles, and is therefore weaker. All day and night, the atmosphere circulates the energy it gets as it is heated by the Sun. Winds such as the monsoon in India and other parts of the tropics, or the Furious Fifties and Screaming Sixties in the Southern Ocean that battered the *Akademik Shokalskiy* (and many other polar expeditions) en route to Antarctica, are the atmosphere's attempt to rebalance the differences in pressure created as certain regions of air become hotter or colder than others and air is compelled to flow from more dense to less dense parts of the atmosphere. As a part of the hydrological cycle, these winds help distribute heat and water from the ocean to the furthest interior reaches of the land. In the process, they make much more of the Earth habitable than it would otherwise be.

At the local level, winds can seem wildly variable in direction and strength. But step back and look at the entire globe, and you will see distinct patterns of circulating winds arranged in belts around the Earth. These global weather systems are so reliable that explorers and armies have used them throughout history to help launch empires.

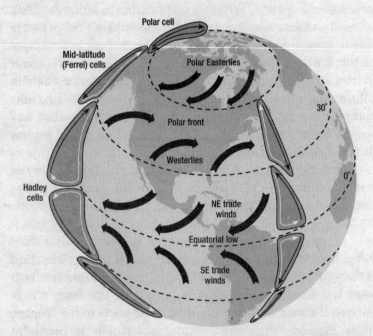

European ships filled their sails with the Trade Winds, which blow continually from east to west in the tropics, and used that energy to open up their continent to the new world of the Americas in the fifteenth century and beyond. We now know that these winds are part of a system of convection currents in the atmosphere lined up around the equator.

The currents, called atmospheric cells, closest to the equator are named after the lawyer and amateur meteorologist George Hadley, who had set out in the eighteenth century to better understand and predict the cause of the winds, at a critical time when shipping to Asia and the newly discovered Americas was getting ever busier.

Hadley cells form either side of the equator when warm, moist air is heated by the Sun. Here it rises, travels towards the poles and, at around 30 degrees latitude in each hemisphere, falls back to the surface. They are not the only atmospheric cells – further north in the northern hemisphere (and south in the southern hemisphere), are another belt of winds named after the nineteenth-century meteorologist who first proposed them, William Ferrel. Furthest from the equator in each hemisphere, blowing above the coldest reaches of the icebound lands, are the winds that make up the polar cells. Each set of atmospheric cells has specific winds associated with it, reliable weather systems that we have come to measure and understand over centuries of travel. Most importantly, the winds do not only move water vapour through the troposphere, they also drive the currents on the surface of the ocean itself.

More than a decade before he became one of the founding fathers of the United States of America, Benjamin Franklin was working on a conundrum that had been brought to his attention about ships crossing the Atlantic Ocean. Packet ships carrying mail from Falmouth, which is near the Atlantic coast at the south-western tip of England, to New York took two weeks longer to make the journey than merchant ships travelling from London to Rhode Island. The distance from London to the east coast of northern America was clearly greater than it was from Falmouth; and the distance from Rhode Island to New York was barely a day by road. So why the strange discrepancy in the ships' voyage time? This caught Franklin's attention because, at the time in 1769, he was the Deputy Postmaster General in North America and delays in sending mail were his problem.

Franklin relayed the puzzle to the captain of a Nantucket whaling ship whom he knew, Timothy Folger, who happened to be in London at the time. Folger immediately knew where the answer lay. He told Franklin that the difference was down to the fact that the Rhode Island merchant ship captains knew about a strong water current in the Atlantic Ocean, which flowed in certain parts of the sea from west to east, while the captains of the English packets ships did not. The American captains knew about it because they noticed how the whales they chased tended to stay near the sides of the current but never went into it. They also knew something was different there because they saw that the water in the current was a different colour and flowed faster. Sometimes, Folger told Franklin, when the whaling ships crossed the current to get on the other side of the whales, they would run into packet ships that were plugging their way against the strong flow of the water. The whalers would advise the packet ship crews to get out of the current. But, as Franklin wrote in his recollection of his conversation with Folger, the English captains 'were too wise to be counselled by simple American fishermen'.

Franklin and his whaling friends had not discovered this current themselves. In 1513, Juan Ponce de León, Spanish explorer and first governor of Puerto Rico, noticed his ships getting into trouble fighting against a strong current next to the coast of North America during the first European expedition to a land that he later named Florida. But Franklin, when he found out about the current, asked Folger to mark out its borders and direction of flow. Along with temperature and current measurements he made himself on later journeys, he published the first map of the current in 1770 and proposed a name for it: the Gulf Stream. A vessel travelling from Europe to North America could shorten her passage by avoiding sailing against the stream, he wrote, while 'a vessel from America to Europe may do the same by the same means of keeping in it. It may have often happened accidentally, that voyages have been shortened by these circumstances. It is well to have the command of them.'

The Gulf Stream is a current of deep blue warm water that moves at up to 4mph as it snakes its way from the Gulf of Mexico, up along the eastern coast of North America, round to the British Isles and then towards Iceland and beyond. Moving along the surface of the water, this current warms Florida and gives the residents of the west coast of Ireland, England and Scotland more tolerable weather when compared to other countries on similar latitudes in, say, Canada or Russia. On the west coast of Norway, the Gulf Stream's winds and warm waters keep most of the coast ice-free for large parts of the year. And it moves water at an astonishing rate – 30 million cubic metres of water flow through it per second as it pours through the straits of Florida, some fifty times more than the combined flow of the mighty Mississippi, Amazon and other rivers that pour out into the Atlantic Ocean.

The path of this current is governed by the winds blowing at the surface of the water. Because the air at the equator is in direct sunshine, it gets hot and sets up Hadley atmospheric cells that move winds north and south towards the cooler poles. While these winds make their way from the equator, the Earth meanwhile rotates from west to east underneath the atmosphere. In the northern hemisphere, this 'Coriolis effect' makes the winds appear to curve to the right as they move north. The opposite happens in the southern hemisphere, where the winds curve to their left as they head south. In the northern Atlantic Ocean, the result is a clockwise stream of air that moves warm air north and eventually returns cool, polar air towards the equator.

As this wind moves around the ocean, it drags the surface of the water with it, creating a gyre in the upper reaches of the sea. The Gulf Stream is part of a system of wind-driven currents that form the North Atlantic Gyre, which starts when water is driven westward from the east coast of Africa by the Trade Winds. By the time the Gulf Stream's waters move around the gyre to northern Europe, it will have split into a multitude of swirling currents and eddies at the surface of the water. One of these branches, the North Atlantic Current, moves north to the UK and Norway, while the Canary Current takes water back towards the equator at the eastern edge of the Atlantic.

There are five wind-driven gyres in the world's oceans: one each in the north and south Atlantic and Pacific Oceans and one in the Indian Ocean. In the everyday maelstrom of the oceans, the gyres circulate water like clockwork. Drop a message in a bottle into the North Pacific Gyre off the coast of California, say, and it will reliably come back to you in around six and a half years after a 14,000 nautical-mile journey. Around the Atlantic gyres, you can expect the return of your bottle in around three years.

One of the modern, unfortunate results of these strong ocean circulations is the collection within them of non-biodegradable materials such as plastic. In the North Pacific Gyre, this has become such a problem that a large area within the gyre has been nicknamed the 'Great Pacific Garbage Patch' by scientists. In certain parts of the sea here, tiny particles of plastic (waste from our modern lives) outnumber the phytoplankton. Their overall effect on the ocean is not yet known – even if they are not a problem for the ocean itself, the hormone-mimicking chemicals many artificial materials contain (known as 'endocrine disrupters' because they can be absorbed and concentrated into animals and cause growth problems) could end up leaching out and causing problems for fish and other species that live and

OCEAN GYRES

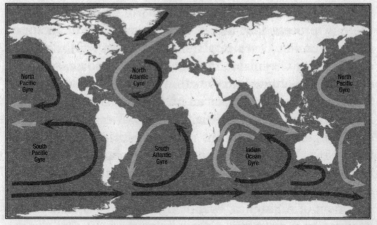

breed in the area. Other things also find themselves trapped in the gyres that spin around seemingly for aeons: cargo containers lost at sea (which go unreported), dead bodies and even entire ships that have been abandoned, left to circle the ocean's waves forever.

These surface currents, driven by wind, are the movements that affect human life most directly. Ship captains need to know how these currents are behaving on a daily and weekly basis to work out the best routes between continents. Fishermen use them to track their catches. People who want to sail need to know how and where is safe to travel. Swimmers and surfers need to know about coastal tides and currents to avoid drowning. But, as useful and important as they are, these surface water currents only affect the upper 400m or so of the world's oceans, around 10 per cent of the water that is contained there. The remainder, the deep water, has its own set of currents that move energy and nutrients across the entire globe in timescales that defy any human imagination.

After the Gulf Stream has crossed the Atlantic to Europe, the part of it that snakes past the British Isles and Norway will move onwards to the frigid polar regions. Some of the water at the surface of that current will freeze once it gets there, a process that drives out the salt present in there, leaving the liquid sea remaining underneath it saltier and denser. Near Greenland, the water becomes so dense that it sinks to the bottom, into something known ominously as the 'abyssal ocean'. That begins an underwater river, deep in the ocean, called the North Atlantic Deep Water current. This cold, languid stream creeps its way south at the bottom of the ocean at less than a centimetre per second. After hundreds of years it reaches the southern tip of South America, where it joins another cold underwater current that circles the frozen continent of Antarctica. Eventually, the water flowing in this deep-water current circles the whole world.

Around 90 per cent of the water in the oceans sits below 400m, and this is the world of the deep ocean currents. Propelled by gravity, denser water sinks and less dense water rises up. The density of a specific parcel of sea water depends on its temperature and saltiness. Places where there is an excess of cold, salty water flow downwards and are known as 'sinks'. Anywhere that is warmer and less salty, the water tends to flow to the surface in what is called an 'upwelling'.

The biggest sinks on Earth are in the North Atlantic in the Labrador and Greenland seas where, as we have seen, parts of the Gulf Stream sink into the deep ocean. At several places around Antarctica, the cold, salty surface water sinks in the same way and there are four main areas where Antarctic Bottom Water is formed: the Weddell Sea, Prydz Bay, Adélie Land and the Ross Sea. At these places, cold polar winds blow across the water and freeze the surface, leaving the water directly underneath saltier and sinking into the abyss. There it joins the deep-water current flowing around the Antarctic and then on to the warmer seas of the equator and beyond.

Put all of these deep-water currents together and you

get the enormous, Earth-wide system known as the Global Conveyor Belt, driven by the thermohaline circulation ('thermo' meaning heat and 'haline' meaning salt). A molecule of water will take more than a thousand years to move around the conveyor belt and the circulation is an important part of the way nutrients, energy and carbon dioxide move around the ocean. The warm waters at the surface of the ocean often become depleted of the nutrients that organisms at the base of the marine food chain – algae and seaweed – need to thrive. The churn from the cool waters of the languid conveyor belt can recharge those surface waters at upwellings and keep the oceans churned and fit for life.

THERMOHALINE CIRCULATION

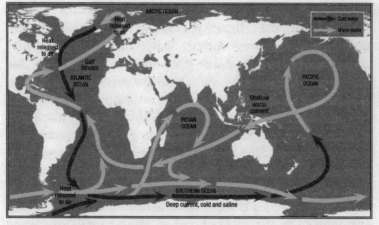

Given all this movement and mixing, climate scientist Carl Wunsch says that the ocean is best viewed as 'a mechanically driven fluid engine, capable of importing, exporting, and transporting vast quantities of heat and fresh water'. The whole system operates like a giant conveyor belt that only works because it follows the laws of physics, moving salt and cold water (and any man-made rubbish) towards the equator via deep currents while warm water moves towards the poles at the surface.

The present system of surface and deep-water currents has not been there forever, however. As the conditions on Earth have changed, so have the way the air and water moves. During the last Ice Age some 12,000 years ago, the deep-water parts of the current carried less water than they do today and the sinking of water in the polar regions looks like it might have slowed to a crawl or even stopped for a while.

Evidence from the chemicals in the fossilised shells of a deep-sea creature called foraminifera show that around 55 million years ago the deep-water circulation reversed abruptly, probably caused by a bout of global warming due to increased greenhouse gases in the atmosphere. Around this time, the Earth was in the Palaeocene/Eocene Thermal Maximum (PETM) climate period and the average temperature of the oceans was some 7 or 8°C higher than today.

Oceanographers at the University of California, San Diego, working on the switch in deep-water circulation, looked at the relative amounts of two isotopes of carbon in the shells of the foraminifera. As water travels along the deep-water current towards the equator, it becomes enriched in carbon-12 relative to carbon-13 because it collects more dead organic matter from sea creatures that die at the surface and fall to the bottom of the ocean. As a rule, living things on Earth tend to use more carbon-12 to grow, and looking at a marine fossil's ratio of carbon 12 to 13 will give you a good indication of where it is from, in other words how far from the equator it died and fell to the sea floor.

By comparing the relative composition of carbon in 55-million-year-old foraminifera shells at different places in the Pacific and Atlantic Oceans, the oceanographers saw that the gradient in relative amounts of the two types of carbon was reversed – the Atlantic's deep-water current was flowing from north to south. Intriguingly, the flip in direction seemed to occur very fast – in just a few thousand years – but took more than 100,000 years to revert.

It is not clear what caused these huge changes, but it is likely that the dense, salty water near the poles would have had to become fresher in order for the current to become compromised. Perhaps the warmer seas produced more evaporation in the tropics, which then turned into clouds and rained down at higher latitudes, dumping more fresh water into (and therefore compromising the action of) a part of the ocean that would normally be a sink, where the dense, salty water drives one end of the vast global conveyor belt. We don't know exactly what happened but we know that these changes to climate can be fast and huge.

This is a stark warning for our future. The oceans continue to play a central role in the Earth's life and climate. Marine life produces half of the oxygen in the atmosphere through photosynthesis and provides a fifth of the animal protein eaten by more than 1.5 billion people. The oceans are a place of rich biodiversity. They are also essential to modern global industry: 90 per cent of the goods used by the people are shipped across them. But our planet's history shows that this critical resource is remarkably fragile.

Our use of fossil fuels since the Industrial Revolution, in particular, has wrought important changes, both physical and biological, to all parts of the ocean. It is not an exaggeration to say that we are engaged in a huge experiment with the Earth as part of our need to pump ever more greenhouse gases into the atmosphere and watch the world warm. This is not a book about climate change but I do want to consider some of the key physical and biological impacts that our actions are having on our oceans. It would be wilfully blind to do otherwise.

*

One of the key indicators of climate change is changing temperatures in the world's oceans. One hundred and thirty-five years after the Challenger Expedition made the first systematic measurements of the world's ocean temperatures and found how different layers of the ocean had different temperatures, oceanographers Dean Roemmich at the Scripps Oceanography Institute in California and John Gould at the National Oceanography Centre in Southampton decided to use modern equipment to retrace the path of the HMS *Challenger*.

They wanted to measure the current temperature of the ocean at the same points and depths that Charles Wyville Thomson's scientists had done in the nineteenth century. But Roemmich and Gould did not need to finance an expensive new expedition to make their comparison, nor did they spend years travelling the world's oceans and lowering thermometers into the water. Instead they used data from a network of robotic floats that had been drifting around the world's oceans since 2004, called Argo. Named after the mythical Greek ship from which sailors chased the golden fleece, more than 3,000 robotic Argo floats now wander the world's ocean currents, each one bristling with temperature, salinity, current and wind sensors. They descend to up to 2km below the surface of the water and near-continuously measure the water's physical properties. Every few days, the floats rise to the surface and upload their position and measurements to oceanographers via a satellite (which is, appropriately, called Jason after the leader of the Argonauts). The oceanographers on our expedition aboard the *Akademik Shokalskiy* had made their own contribution to the growing Argo network by dropping a float into the Southern Ocean (see Chapter Two) just a few days into our journey to Antarctica.

Because there are so many Argo floats travelling the oceans, Roemmich and Gould were able to recreate the winding route of the HMS *Challenger* by picking data from specific floats that had passed into the 300 or so locations around the world that had been sampled by Thomson's scientists 135 years earlier. Where possible, they even selected Argo data from the same time of year as when Thomson's team would have been sampling, to get the best possible comparison between old and new measurements.

They found that, on average, the surface of the world's oceans had warmed by 0.59°C between the Challenger Expedition of 1872–76 and the Argo measurements of 2012. Down to 700m from the surface of the ocean (around 400 fathoms), the average warming was 0.33°C. Average warming on the surface of the Atlantic Ocean (1°C) was stronger than that in the Pacific (0.41°C). Increasing sea surface temperature is important for several reasons. This is where the interchange of energy between the oceans and the atmosphere happens, a key piece of the global hydro-logical cycle. A higher sea surface temperature means more evaporation from the oceans and, therefore, more precipita-tion (and inevitably bigger storms) somewhere else in the world. It also means sea-level rise – due to the water expanding as it warms and the addition of water from melting glaciers and ice sheets – and thus greater inunda-tion and flooding of coastal cities and habitats.

Because of systematic limitations in the way that *Challenger* made its measurements in the 1870s, Roemmich and Gould wrote in *Nature* that the temperature changes they had found since that expedition were probably on the conservative side, with the real changes likely to be greater. 'This study underlines the scientific significance of the *Challenger* expedition and the modern Argo Programme,' they wrote, 'and indicates that globally the oceans have been warming at least since the late-nineteenth or early twentieth century.'

*

That the sea is warming should be of no surprise to anyone who has watched the growing drip-feed of scientific evidence in recent years about the physical health of our planet.

In 2014, the Intergovernmental Panel on Climate Change (IPCC) published its fifth assessment of the evidence around the effects of greenhouse gases in the atmosphere. Their comprehensive review involved thousands of climate scientists who spend years poring over the research literature on everything from atmospheric physics to marine populations. Their thousand-page reports give policymakers (and anyone else who wants to read them) a comprehensive overview of how the climate has changed since the Industrial Revolution, how fast and what effects it has already had. In addition, the IPCC looks to the future and outlines scenarios for what will happen to everything from animal populations to habitats to sea levels if we continue to pump greenhouse gases into the atmosphere. Each statement – historical or prediction – comes with a confidence rating to show how well supported it is from the best available evidence.

In its 2014 report, the IPCC incorporated the latest data from the worldwide Argo float network to conclude that our oceans have absorbed 93 per cent of the extra energy from the enhanced greenhouse effect caused by our use of fossil fuels over the past century and approximately 30 per cent of anthropogenic CO_2 from the atmosphere. Sea surface temperatures around the world have increased since the start of the twentieth century, and since 1950, the average rises in the Indian, Atlantic and Pacific Oceans are 0.65, 0.41 and 0.31°C respectively, with most of that warming occurring in the upper 700m of water. Depending on how the coming century unfolds, the seas could be anything from 1–3°C warmer, or more.

Warmer seas power stronger storms in the tropics, with their potential for great damage if they make landfall in populated areas of the world. The IPCC review said it was virtually certain that the frequency and intensity of the strongest tropical cyclones in the North Atlantic had increased since the 1970s.

There are more subtle and, in global biodiversity terms, more significant effects of warmer seas. Higher temperatures mean changes in saltiness and, therefore, ocean dynamics. In some tropical and higher latitudes, the warmer sea means that more rain or snow will fall. And, near the poles, more sea ice will melt. Taken together, these increased sources of fresh water will reduce the saltiness of the sea at high latitudes and therefore compromise the production of the dense, salty water that normally sinks near poles to drive the Global Conveyor Belt. A slowdown in this current could, in the long term, have far-reaching global consequences for weather and climate.

Closer in time, this change in ocean dynamics will also affect biodiversity. The ocean is stratified into layers of different densities, which are determined by their temperature and amount of salt dissolved in them. Warming of the ocean surface and an influx of fresh water (from ice melt or more rain or snow) can exacerbate those differences in density and create even more stable layers. That is bad news for life at the surface – the warmer, less saline surface layers tend to contain a lower concentration of nutrients than the colder, more saline layers underneath. Normally the layers are whipped up by winds and mixed together, allowing nutrients to flow between them. This mixing, though, is less likely to happen as the water densities get further apart. A consequence of this is the creation of 'oxygen minimum zones' (colloquially called 'dead zones') in the ocean, where certain organisms deplete the limited dissolved oxygen in that layer which, because of the lack of mixing, does not get replenished. This leaves larger fish and mammals unable to survive there.

Dead zones can also be caused by pollution and runoff from agricultural land. If too many nutrients and fertilisers wash off land, either into rivers near farms or directly into the sea from coastal cities, they can stimulate an excess growth of phytoplankton and algae. Though this does sequester an extra amount of carbon dioxide from the atmosphere, another result is that this bloom sucks out most of the oxygen in the water. Warmer waters can also stimulate the production of blooms.

These are only a fraction of the changes that have occurred in the world's oceans. They have happened relatively quickly, in around 150 years. But reversing them will take our planet an extremely long time, perhaps thousands of years according to the IPCC.

Because of the large mass and heat-absorbing capacity of the water within oceans, they can absorb a thousand times more heat than the atmosphere. From 1971 to 2010, the oceans have, as we have seen, taken in more than 90 per cent of the Earth's heat gain due to climate change. To reverse that heating, the warmer upper layers of the ocean will have to mix with the colder deeper layers and that process could take up to another thousand years. This means that it will take centuries, perhaps millennia, for the deep ocean to warm in response to today's surface warming, and at least as long again for the ocean warming to reverse if we manage to reduce our atmospheric concentrations of greenhouse gases.

All the extra CO_2 in the atmosphere has also made the water more acidic. This gas dissolves in water to form carbonic acid and this has led to the ocean's pH decreasing on average in the past century by 0.1 units. The IPCC assessment is that the current rate of ocean acidification is unprecedented within the last 65 to 300 million years.

Animals that live in the water tend to have up to twenty times less CO_2 dissolved in their blood than terrestrial creatures, so just a slight change in the amount of this substance in the water will have bigger relative impacts on marine animals.

Many marine species, including plankton, shellfish and coral, build their shells from calcium carbonate. Sitting in more acidic waters is bad enough, since this water is better at damaging and dissolving the shells. But it also causes problems in making the shell in the first place. Shell-making organisms get the calcium they need from the sea water around them but to turn it into shell material the organism needs to bring the calcium into specific places in their bodies that are more alkaline (higher pH) than the rest of their bodies and also the sea water around them. If the sea water around them is getting increasingly acidic, it will take the organism more energy to make the calcification process work, since it will need to create a bigger differential between the outside and inside of its body. The more energy it spends on building its shell, the less it has for everything else, such as growth or reproduction. In this environment, the organism will then have a disadvantage and get into a vicious cycle where it is less able to compete effectively for food or other resources, therefore making it increasingly weaker. This is the recipe for a population collapse.

Ocean acidification has put additional pressure onto some of the most iconic marine organisms, coral, by degrading their shells and reefs. Corals reefs are the Earth's most diverse marine ecosystem, often referred to as the rain forests of the oceans. They take up less than 0.2 per cent of the ocean floor, yet they host up to a third of all marine species, providing both food and shelter. The marine ecosystems around coral were already having problems due to warmer waters, pollution, invasive predators and increasing storm damage. Rising sea temperatures cause corals to bleach, a process where the coral tissues expel their symbiotic algae, called zooxanthellae. These algae not only give the coral their colour, but allow them to gain food from photosynthesis. Without them, the corals will starve to death within a few months. Recent long-term surveys have shown that the Caribbean's corals have declined by more than 50 per cent since 1970 and could disappear completely within twenty years; the Great Barrier Reef's coral cover has dropped by half since 1985 and is predicted to halve again by 2022. These changes to the ocean's carbonate chemistry will take tens of thousands of years to reverse.

'Climate change alters physical, chemical and biological properties of the ocean,' said the IPCC. Salinity, circulation, temperature, carbon dioxide, oxygen, nutrients and availability of light will all change as a result and these, in turn, will determine the success or failure of marine organisms from phytoplankton to fish to blue whales.

Warming temperature, falling pH and changing salinity are just a few of the many physical effects we can see on the oceans. Their impacts reach deep into marine and coastal ecosystems that have evolved to live in specific niches (and, via the ocean's effects on weather, ecosystems on land too). Much of this rapid change can already be seen with marine organisms such as fish and plankton moving to higher latitudes so that they can remain in an environment more suited to their evolution, and changes in the timings of breeding and migration. 'There is medium to high agreement that these changes pose significant uncertainties and risks to fisheries, aquaculture and other coastal activities,' wrote the IPCC.

Whereas the production of phytoplankton will likely increase at the higher latitudes (because the amount of available sunlight for photosynthesis increases as there is more uncovered ocean because of melting ice), the phytoplankton at mid-latitudes and the tropics will decrease because of a lack of nutrients. Because these organisms are at the base of the food chain, their abundance has important effects for all the species further up the ladder, from invertebrates to fish to whales. Most climate change scenarios predict a shift or expansion of the ranges of many species of plankton, fish and invertebrates towards higher latitudes, by tens of kilometres per decade.

This will have important impacts for fisheries – those in the tropics will see a 40–60 per cent drop in their yield by 2055, when compared with 2005 and based on a 2°C average temperature rise above levels at the Industrial Revolution, before humans started pumping copious CO_2 into the atmosphere. At mid to high latitudes, it's a different story and fishermen can expect a 30–70 per cent increase in what they catch in the same timeframe.

In the polar regions, the marine species adapted to the cold waters will be squeezed as they will increasingly have nowhere to go when the water around them warms. Species in semi-enclosed seas such as the Wadden Sea and the Mediterranean Sea, for example, face a higher risk of local extinction because land boundaries around those seas will make it difficult for them to escape waters that may become too warm.

The ocean floor (benthic) regions might not have the same level of physical changes in terms of temperature or acidity, but they will nonetheless be affected by the knock-on effects of the open ocean (pelagic) changes. Even in the deepest abysses, living things depend on a continuous rain of dead organic material from the surface for food.

Perhaps the most talked-about effect of climate change is sea-level rise and with good reason – the rate of rise since the mid-nineteenth century has been larger than the average rate during the previous two millennia. From the period 1901 to 2010, the world's average sea level rose by 0.19m, at a rate of 1.7mm per year between 1901 and 2010, 2.0mm per year between 1971 and 2010 and 3.2mm per year between 1993 and 2010. Different parts of the world have seen (and will continue to see) different amounts of sea-level rise – rates are three times higher than the average in the Western Pacific and South-East Asian region, whereas they are decreasing in the Eastern Pacific for the period 1993–2012.

This change in sea levels will continue for hundreds of years as we put more greenhouse gases into the atmosphere. The projected sea-level rise by the end of this century is anything up to half a metre on average, depending on the scenario and how much greenhouse gas ends up in the atmosphere by then. Rising seas have many adverse consequences, not least vast changes to coastline ecosystems such as beaches, salt marshes, coral reefs and mangroves. People living near coasts will also face increased flooding and more frequent extreme weather events, leading to damage and destruction for people living in coastal communities.

'Evidence that human activities are fundamentally changing the ocean is virtually certain,' wrote the IPCC. Sea temperatures have increased rapidly over the past 60 years, the seas have become more acidic and the rate at which these fundamental physical and chemical parameters of the ocean are changing is unprecedented within the last 65 to 300 million years. Not surprisingly, these fundamental changes will lead to negative impacts on the hundreds of millions of people that depend on the ocean ecosystem for food, shelter and other services.

Species do adapt and evolve to changing climactic conditions on Earth and, in its 4-billion-year history, our planet has seen much bigger overall climactic differences than the ones I have described above. The difference now is the pace of change, something that no life form can easily cope with given the languid, random nature of evolution. The resilience of marine species to cope with the changing water currents, temperatures, chemical and biological composition is severely limited. 'The current rate of environmental change is much faster than most climate changes in the Earth's history, so predictions from longer-term geological records may not be applicable if the changes occur within a few generations of a species,' concluded the IPCC. 'A species that had more time to adapt in the past may simply not have time to adapt under future climate change.'

In 2008, a group of artisan miners in western Brazil found a tiny, battered-looking diamond in the gravel of a river bed in Mato Grosso. It looked more like a nugget of dull metal than a gemstone and had pits and scratches across its surface, the marks of its violent journey being blasted onto the surface of the Earth from a depth of 500km by a volcanic eruption.

The more beautiful diamonds prized for jewellery come from shallower depths, around 150km, but the ugly Brazilian gemstone from Juína contained a more significant cargo than anything suitable for an engagement ring. Invisible to the naked eye, inside the 5mm-long stone was a clue to something profound about our water world, a minuscule indicator about how little we really know about how much of this substance there is on our planet.

In this chapter, we have discussed the water present on the surface and in the atmosphere of our planet. This is the stuff that is most important to us, our fellow living things and also to the shape of the planet as we know it. The hydrological cycle and its components have carved continents and allowed some areas to bloom with vegetation while leaving others bare, exposed to the elements.

But the diamond from Juína in Brazil shows us that there is more water on Earth. Much more. This stone demonstrated just how little time we have spent probing underneath the skin of our planet. Inside the diamond was an inclusion of a water-rich mineral called ringwoodite, a form of the magnesium-silicon compound olivine that forms under extreme pressures. Scientists had long suspected that ringwoodite made up large portions of the Earth's mantle but, without this shabby diamond messenger, they would not have been able to confirm it. Their oblong stone contained around 1.5 per cent water by weight, which doesn't sound like much but, given the vast amount of ringwoodite that is thought to exist in the Earth's crust, it could mean that there is a 'wet zone' deep inside the planet that holds as much water as all of the world's oceans combined. It would not be in the form of a liquid sea but, instead, bonded within high-pressure minerals such as ringwoodite or wadsleyite (both of which can contain up to 2.5 per cent of their weight in the form of water) inside the transition zone of the Earth's mantle, which stretches from 400km to 650km under the surface.

We know that Earth is a water world and this is usually an ostensible reference to the great oceans, rivers and clouds that we can see. But the material of the world itself, the rocks and the dust, are also drenched in more water than we could ever have imagined.

Part II

⊷

BIOSPHERE

FIVE

Life

'We start with a single, innocent looking, apparently very simple water molecule. One oxygen atom attached to two hydrogen atoms. It could hardly be simpler. Yet this simple entity is the ultimate culprit behind all the beauty and complexity. Let's start from here and see how such a small, straightforward molecule can give rise to such world-shattering – and world-stabilising – complexities.'

John Finney, 2004

On board the *Akademik Shokalskiy*, scientists were looking for living things in the ocean at all possible scales. A team of mammal biologists spent their time on the lookout for Weddell seals, leopard seals and (fingers crossed) whales. They would peer through binoculars from the sides of the ship, listen for underwater songs with hydrophones and sometimes go out onto the water itself with waterproof speakers to play songs under the waves, hoping to lure creatures towards us.

Ornithologists took up posts on the bridge to count, every hour, the birds they could see – their passage of latitudes as we sailed south were marked by the changing populations of albatrosses, skuas, petrels, shearwaters and fulmars that swooped by our rolling ship. These open-ocean birds soared over the cold waters, sometimes fighting the intense winds, occasionally skimming the water to feed and, rarely it seemed, resting on the tiny flecks of land in the sub-Antarctics in which to nest. We all looked out for penguins – gentoo, royal, rockhopper, chinstrap and, later in our journey, Adélie and emperor.

Marine biologists worked with the oceanographers to collect the microscopic life within the water itself – their fine-gauge nets dragged behind the ship and caught a lumpy green soup of plankton, a record of the changing biodiversity along the path of the journey. As unassuming and uncharismatic as the sludge was, it held an importance for all the animals we saw from the ship. This microscopic world, embedded in the fluctuating fabric of the Southern Ocean itself, is where the biology of this great ecosystem begins.

The complex, floating ecological web of the ocean is powered, like everything else on Earth, by sunlight hitting the surface of the world. Here, across the thousands of square miles and down to the first few hundred feet from the surface of the ocean, is the primary entry-point of energy into biology.

In every droplet of water in the ocean, there is life. The surface layer (the euphotic zone) blooms with tiny plant-like organisms, phytoplankton. These needles, discs, balls, spurs and rhomboids photosynthesise sunlight with the carbon dioxide dissolved in water to create sugars and other organic materials. Invisible to the naked eye, they nevertheless account for half of the photosynthesis happening on our planet.

The phytoplankton are a multitude of species of plant-like cells that use chlorophyll and other pigments to harness sunlight and convert the carbon dioxide dissolved in the sea water into organic compounds that they use to build their bodies. They drift with the ocean currents and bloom every summer near the surface as the sunlight intensity increases.

In places where there are plenty of nutrients, the blooms can turn patches of the winter-cold, clear water into a dense soup of suspended organisms.

The most common, diatoms, are encased in silica and come in an array of shapes – glistening diamonds, spikes and translucent pillboxes – and float either in clouds of single cells or form sparkling chains. Some species are buffeted by currents, others have evolved flagella that can help them move as they need. Some phytoplankton live in colonies, loosely bound in a jelly-like matrix. White cocco-lithophores use calcium carbonate in the water to build tiny chalk plates, each one looking like a warrior's shield, which they use to pack around themselves for protection. These plankton can swim and some of them contain droplets of oil to help them stay afloat. Blooms of them can turn the water milky, a coloured cloud of water that can often be seen from space. When they die, the remains of the cocco-lithophores, each one an invisible speck in the water, fall to the bottom of the ocean. Over the course of millions of years, these carcasses create deep sediments of chalk. Some of the sediments that formed in the cretaceous period became what we now see as the vast chalk cliffs facing the sea in the downs of the south east of England.

Phytoplankton are the leaves of the ocean, distributed and roiling through the waves, using sunlight to create organic matter, turning mere physics into our biology. In turn, they are consumed by companion organisms in the water, tiny animals that drift among the plant-like cells, the zooplankton. These herbivorous grazers include copepods, creatures the size of a grain of rice with external skeletons – the marine equivalent of insects. Several thousand species of these feast on the phytoplankton, including crustaceans such as the long, shrimp-like krill. These move with the currents, catching what phytoplankton they can and living and dying on the tides.

The copepods are eaten by the next animal up the zooplankton chain, the amphipods, which are altogether more complex, with eyes, multiple jointed legs and hunting strategies evolved to maximise their intake of food. Arrowworms have bristle-like jaws, sea gooseberries use retractable tentacles, sea angels and sea butterflies use modified feet to move and chase their prey and the latter even creates a 'net' using mucus. Some are gelatinous tubes that look like naked slugs while others resemble miniature woodlice, gnats or fleas, with segmented bodies and ridges along their tops, and coloured in a combination of reds, browns, oranges, yellows and creams. Other species, including worms, light up the water around them with chemical reactions inside their otherwise transparent bodies.

In our droplet of ocean water we would also find countless larvae – newborn crabs, comb jellies, shrimp, urchins, cockles, snails, worms and starfish – making their first attempts at survival in the vast ocean. They eat what they can and hope they don't get eaten by something bigger. Crab larvae, for example, come with spines and spikes to try to stay alive. All these larvae are only temporary plankton, however, and when they are big enough they drift down to the muddy sea bed to live out their adult lives.

Invisible to us, each droplet of ocean water is a living world.

All of the life on Earth – from archaea to axolotls, bacteria to blue whales – is made mostly of water. Every dynamic process in cells, every biochemical reaction needed to keep things alive, moving and growing, depends on water. If we think of life on Earth as a vast series of trades and transactions, water is the currency in which these exchanges are made.

The metabolic processes inside cells – to build proteins, for example, or generate useable fuel to keep the body running – require either the addition or removal of water molecules. Plants and many bacteria use the Sun's energy to split water in the most important chemical reaction on Earth – photosynthesis. Once split, hydrogen atoms are combined with carbon dioxide to produce glucose, and oxygen is released into the air. This reaction ultimately provides food for all other living things. Cells turn glucose back into water and carbon dioxide, releasing the energy they need to power cellular functions in the process.

To say water is integral to life understates the case. We should properly think of our cells – and those of every living thing on Earth – as bubbles of water that contain tiny amounts of carbon, hydrogen, oxygen, nitrogen, phosphorus and sulphur, suspended or dissolved inside them. Life, its processes and structures, occur in its solution. Every living thing, then, is simply a different inflection of water; a deviation of a few per cent from purity.

We often hear life on our planet described as being 'carbon-based'. It's true that carbon's ability to form long chains with itself and other key elements provides the bricks of life – complex molecules such as DNA, proteins and lipids. Water, however, is the cradle in which life can happen. It is the crane, the scaffolding and mortar of existence.

Water moves things around cells and bodies, keeps chemical reactions going, cleans and takes away waste, maintains temperatures, lubricates, acts as a shock absorber and a building material. Its strange chemistry is important too – water's large heat capacity allows water-filled organisms to manage external temperature fluctuations that might otherwise kill them. Cells may well survive if their water content gets drastically low but they stop working unless they have the right amount of hydration.

In all of this, water is no background player, a mere arena of action, but an active participant and biomolecule in its own right. Without the liquid medium of water in which to exist, the complex life molecules we know of – proteins, DNA, sugars, salts and lipids – would be nothing more than chemicals, ingredients that we could not define as biological.

Across all domains of life, the amount of water contained within varies from just shy of 50 per cent (in some bacterial spores) to a whopping 97 per cent (in some marine invertebrates). Adult humans consist of around 60–70 per cent water on average, though it is not uniformly distributed around the body – the nervous system's tissues are 84 per cent water, the liver is 73 per cent, muscle is 77 per cent, skin is 71 per cent and fat is around 30 per cent. Fluids such as blood plasma (the part that isn't blood cells), saliva and gastric juice are almost entirely water. The amount of water in a person changes during developmental stage, especially early on. A foetus in its first months is around 95 per cent water and gets to 77 per cent by the time it is born, with a child only reaching the 60–70 per cent level by the time he or she has 'matured' half-way through the first decade of life and gained more muscle and other minerals. That remains about constant and then the water content starts to diminish in older age.

Two-thirds of the water in your body is inside cells, which means that a 70kg person is made up of around 42 litres of water, of which 28 litres is the water in their cells. Extracellular water is contained in blood plasma (3 litres) and the gaps between cells and in lymph (10 litres). The remaining litre of water is distributed in fluid contained in the spine, is the stuff that keeps our eyes at the correct shape and pressure, surrounds the abdominal organs and fills and lubricates joints. The body can use the last of these as an emergency source of water if it is in dire need, though the effects – softened eyeballs and constipation – are not pleasant.

Every day we need to drink a litre and a half or more of water to maintain healthy function and more comes through food (more than half of 'solid' food is usually water). We also produce around 500ml of water every day in the chemical reactions that liberate energy as we digest that food. It is remarkably difficult to achieve any major change in water levels in the body because of our ability to maintain homeostasis – you will become thirsty if you lose only 1 per cent of your normal water content.

'Molecular life began bathed in it, and has never found a way of getting away from it,' wrote biologist Peter Rand about water. 'Much of evolution has provided mechanisms not only for keeping it there, but also for making sure it does not change concentration, even in the face of drastic environmental conditions.'

The idea that water is a special substance, crucial for life, is an old tale, told across many cultures. The Greek philosophers were not the first to describe it like this but it was one of them, Aristotle, who largely shaped the Western view of water as one of the elements of nature, alongside earth, air and fire. And until 200 years ago, water was still identified like this – an element, something primal and indivisible. The seeming lack of interest in studying it further is perhaps a reflection of how ubiquitous water is, how far it seemed from the most interesting edges of knowledge throughout history. Certainly, it indicates the continued hold that Aristotelian thinking had on thinkers and scientists until well into the Enlightenment.

Three men challenged the elemental description of water in the final decades of the eighteenth century, taking the first steps to identifying and writing down the most famous chemical symbol on Earth: H_2O. The credit for the discovery of the structure of water, the last of the ancient elements to be penetrated by scientific techniques, was highly sought after, and the arguments over primacy, later known as the Victorian Water Controversy, raged for more than seventy years. This was about more than just the attribution of a discovery – the events unfolded as the scientific method was finding its feet, science itself was starting to become more professional, and academies and universities were beginning to attempt to direct (and decide who could be involved in) intellectual progress.

The aristocratic Henry Cavendish was a painfully shy man, and a brilliant natural philosopher. One of the so-called 'pneumatic chemists', he, along with contemporaries Joseph Priestley, Daniel Rutherford and Joseph Black, was fascinated by the nature of air. These early chemists had begun to realise that air was not a single thing but a mixture of several types of 'airs', each with specific characters. They believed that common air, the familiar and everyday stuff we breathe, existed alongside something called phlogiston, an invisible substance that was contained in all inflammable things, and was emitted when they were burned.

Cavendish was the first to recognise that hydrogen – which he knew as 'inflammable air' and liberated by dissolving metals in strong acids – was a distinct chemical element of some sort. Using the prevalent theories of the day, he believed this inflammable air must contain phlogiston. In the early 1780s, he began experimenting with mixtures of inflammable and common air by sealing roughly equal amounts of them into a glass tube and passing electrical sparks through the mixture. The gases in his experiments exploded, but Cavendish was particularly intrigued by the fact that the glass apparatus was coated in a layer of dew afterwards. Careful measurements of the masses of the original air mixture and the reaction products showed that all of the inflammable air and around a fifth of the common air had been turned into the dew – something he later identified as pure water – on the inside of the glass. Cavendish went on to work out that the missing part of the common air was 'dephlogisticated air' (which we now call oxygen – that is, air from which the phlogiston had been removed), a substance that had been discovered by Priestley almost a decade earlier.

In further experiments, Cavendish tinkered with different mixtures of inflammable and dephlogisticated airs (hydrogen and oxygen) until he found that, at a proportion of around two parts of the former to one of the latter, the weight of the air mixture was the same as the weight of the water that came out of the explosive reaction. Cavendish concluded that dephlogisticated air was simply water that had been deprived of phlogiston; and that inflammable air was probably water that had been added to phlogiston. When these two airs combined together, he reasoned, water condensed out. It's not entirely clear whether Cavendish realised that water was a compound of two different elements, or whether he thought it was still a single element by itself, which was part of air. At the time, he stuck to the story that, when inflammable and dephlogisticated airs were combined, water somehow condensed out of the reaction.

Priestley heard about Cavendish's experiments and decided to repeat them. Around the same time, he mentioned the work to the engineer James Watt, who had already proposed the idea that water could be turned into a kind of air if it were heated enough. A self-taught man who had contributed to the birth of the Industrial Revolution with his improvements on Thomas Newcomen's design for the steam engine, Watt interpreted Cavendish's experiments as evidence that water was a compound. 'Water is composed of dephlogisticated and inflammable air, deprived of their latent heat [through an explosion],' he wrote in a letter to Priestley in April 1783 after he heard about Cavendish's work. He wrote up his ideas and asked that Priestley present it to the Royal Society in London.

But there was a problem. When he repeated the experiments himself, Priestley did not come up with the same results as Cavendish, and this cast doubt in Priestley's mind over whether Cavendish's ideas were correct. When Watt found this out, he withdrew his letter from consideration by the Royal Society, pending further investigation from Priestley. (Priestley had not, in fact, found errors in Cavendish's work; rather he had botched his own experiments and managed to confuse himself – and Watt – into thinking Cavendish might have been wrong.)

Meanwhile, the third potential claimant to the title was busy with his own experiments. Antoine Lavoisier was a French tax collector who would become one of the most important figures in developing modern chemistry when he disproved the existence of phlogiston. Lavoisier had obtained water by mixing Cavendish's inflammable air (which he called 'hydrogen') and dephlogisticated air (which he later named 'oxygen') in a closed glass vessel. He published this work in December 1783, which spurred Cavendish on to get a report of his own experiments to be read at the Royal Society a month later. Cavendish proposed a compound structure for water, similar to the idea Watt had suggested to Priestley nine months earlier, but made no mention of Watt in his work.

Watt's supporters were incensed when they read about Cavendish's work and urged the engineer to demand that his original letter to Priestley be read out at the Royal Society. A year after he had initially written it, Watt's proposal that water was a compound of airs was made public and, around the same time, he wrote to a colleague in Bristol that he had 'had the honour, like other great men, to have had my ideas pirated' by Lavoisier in Paris and Cavendish in London. 'The one is a French Financier; and the other a member of the illustrious house of Cavendish, worth above £100,000, and does not spend £1,000 a year,' he wrote. 'Rich men may do mean actions. May you and I always persevere in our integrity, and despise such doings.'

The water controversy had begun. Intermittently over the following seventy years, the argument over who 'discovered' water flared up between groups of learned men aligned with either Cavendish or Watt. Watt's son championed his father's case everywhere he could as evidence that the elder Watt, who died in 1819, was more than 'just' a lowly engineer; that this self-made man was capable of ideas worthy of any of the elite, university-trained natural philosophers. The controversy reached its greatest heights for a few years around 1834, when the natural philosopher and perpetual secretary of the French Academy of Sciences in Paris, François Arago, presented his *Éloge historique de James Watt*, in which he claimed Cavendish and Lavoisier had gone so far as to plagiarise Watt's ideas.

Reverend William Vernon Harcourt responded in defence of Cavendish when making his presidential address to the Birmingham meeting of the British Association for the Advancement of Science in 1839. Harcourt, the son of the Archbishop of York, had developed interests in chemistry and geology and he was a part of an emerging scientific establishment in England. He argued that, despite the timelines of which ideas about water had emerged at which points, it was Cavendish who had had the superior, more philosophical approach to his research, while Watt had been vague and indecisive.

The fight between the Watt and Cavendish camps became a battle between nations (Scotland versus England), backgrounds (working class versus aristocrat) and, perhaps most important, the developing consensus over what could be called 'science' and who were 'scientists'. The Gentlemen of Science, represented by the British Association and the Royal Society among others, sided with Cavendish, who was their representative of the methodical, disciplined approach required by the emerging scientific practice, carefully sanctioned by the academy. Many of these people portrayed Watt, instead, as part of a messier, older and more speculative mindset.

By the time of the Karlsruhe Congress of 1860, where water got its H_2O moniker, the Victorian water controversy had been left aside as scientists engaged with the rapidly developing new science of chemistry. Once the structure had been confirmed, it might have seemed as if the job was done. But, as we know now, chemists had barely begun to understand this strange molecule.

To understand what all this water is doing within us (and the bodies of other animals and plants) beyond flowing as a liquid as part of the bloodstream and streaming out as sweat, we need to dig deeper into a few of the more interesting chemical properties of the water molecule we met in chapter one – its polarity, hydrogen bonding and ability to cluster into vast networks.

Recall that the hydrogens and oxygen in a water molecule are bonded covalently, which means they are held together tightly by sharing their outermost electrons. This is a common type of bond in chemistry and the atoms in most organic (in other words, carbon-based) compounds are bonded in this way. This is how elements turn into compounds, with physical and chemical properties often very different from their component parts – glucose is made from carbon, oxygen and hydrogen, for example, but it neither behaves nor looks anything like any of its covalently bonded ingredients.

On top of that, and probably more important as far as the chemistry of life is concerned, water is a polar molecule, meaning its electric charge is not equally distributed along its length. This polarity allows water molecules to interact with each other via hydrogen bonding, whereby the hydrogen of one water molecule is electrostatically attracted to the oxygen of another. This hydrogen bond is typically temporary and very weak, around 5 per cent of the strength of a standard covalent bond that binds the hydrogen and oxygen within the molecule itself.

Water's polarity makes it a fantastic solvent, able to transport and interact with a whole range of chemicals for life. Any compound that interacts with the hydrogen bonds is known as 'hydrophilic' (water-loving). On the flip side, molecules that are non-polar – and that includes many hydrocarbons such as oil or fats – do not interact well with water and are known as 'hydrophobic' (water-hating). These effects are visible to us every day – you can easily dissolve sugar into your coffee or salt in a boiling pan but you know that oil forms a thin layer on top of a bowl of water, refusing to mix.

The hydrogen bonds can therefore form not only between water molecules but between water and, for example, the oxygen or nitrogen atoms that sit along sections of complex organic molecules such as proteins or DNA. As we will see, these weak bonds and their ability to form and break quickly are at the heart of many of the most important biological processes that go on inside cells.

Hydrogen bonds are what scientists call a 'non-covalent interaction' between molecules. There are other interactions like this between molecules, such as Van der Waals' forces, all of them tenuous links between atoms that are not true chemical bonds but which allow molecules to feel each others' presence. They allow molecules to move in concert or respond to each other without becoming fused together. Together they manage most of the everyday goings-on inside cells and keep things ticking along. The NASA physicist Andrew Pohorille, who has spent a large part of his decades-long career working on definitions for life in order to help his space agency look for it outside Earth, goes so far as to say that these non-covalent interactions can actually help to define a living cell.

To understand what Pohorille means, let's look at what a cell typically does. The functions inside a living cell we would recognise include creating membranes, folding proteins into the correct three-dimensional shape so that they can do their jobs properly, reading and transcribing DNA, carrying out enzyme reactions and shuffling things into and out of cells. All of these processes happen to the ingredients of the cells, which are large, complex biomolecules often made of hundreds of thousands of atoms. They might be carried out by different biomolecules in different cells or in different species, but they all need to happen at temperatures and energies appropriate for the environment in which they exist. Human beings maintain a steady temperature of 37°C and all of our biochemical reactions are tuned to work best at that temperature. That's why a fever that raises temperature or hypothermia that lowers it can be so problematic – the change in temperature disrupts the billions of life-defining chemical reactions going on inside us at every moment.

All that day-to-day stuff has to happen largely without making and breaking full chemical bonds, which can require lots of external energy and involves chemical reactions. Non-covalent interactions, mainly hydrogen bonds, can therefore step in to take over to mediate most of the actions within cells. These interactions are strong enough not to get knocked apart by natural changes in the environment (fluctuations of temperature or the sudden increase in activity required when we need to run away from a predator, say) but not be so strong that the billions of daily functions require huge amounts of energy to carry out or reverse.

Think of a non-covalent interaction, such as a hydrogen bond, like a magnetic latch on a cupboard. When pressed together, the two halves of the magnet will easily keep the cupboard door closed. But the latch is easy to separate when you need to open the cupboard. A covalent chemical bond, in this scenario, would be the equivalent of nailing the cupboard closed – you could still open and close it the door, but doing so would take a lot more effort and you need more energy to do it. It would certainly make the cupboard less useful as a place to temporarily store things.

The job of water, life's solvent, is to enable these tiny non-covalent interactions between biomolecules in cells. Water not only allows the molecules inside a cell to flow around and meet each other where necessary, its chemistry keeps the interactions between biomolecules stable and ensures that the non-covalent interactions all happen at the correct strength. In short, without water, life would fall apart.

And the impact of the hydrogen bond to life goes even further – in fact it was important for life before life even began.

The polarity of water has a deep role in the functions of life. The charge separation between the oxygen and hydrogen gives it a kind of built-in structure, something that chemist Andrea Sella calls an 'internal intelligence'. Water's polarity allows chemical ingredients, with no outside direction, to create the structure and processes necessary to become life.

In 1985, American biochemist David Deamer showed a dramatic example of just how fundamental this self-organisation must be by experimenting on compounds that he had extracted from the Murchison meteorite, a carbon-rich rock that had fallen one morning sixteen years earlier near the town of the same name in Victoria, Australia. The meteorite was a type of carbonaceous chondrite, known to contain lots of organic compounds such as amino acids and other molecules found in living things.

Deamer took inspiration for his work from a British colleague, Alec Bangham, a doctor and captain in the Royal Army Medical Corps who worked on cell membranes at the Animal Physiology Institute in Cambridge, England. In the early 1960s, Bangham found that ovolecithin – a fatty substance extracted from egg yolks – organised itself into bubbles with multi-layered walls when thrown into water and given a good shake. These bubbles became known as liposomes and Bangham and Deamer studied them together in the 1970s. Specifically, they looked at ways to make them bigger, work that has gone on subsequently to be useful for pharmaceutical companies who want to use liposomes to carry drugs and genes into the body.

While working together, Bangham and Deamer wondered if these self-organising bubbles might have been the first containers for living things, billions of years ago at the dawn of life. Until then, it had been assumed that cell membranes were made like everything else in the body, from proteins coded by DNA. Of course, that implied that the genes existed before fully formed cells in the history of life. Instead, Bangham and Deamer thought that if fatty molecules existed in the primordial soup – the pools of mineral-rich water that formed on the Earth's surface in its early years – they might have spontaneously formed membranes, which later housed and protected the earliest replicating molecules that went on to form life.

Deamer took the idea further by recreating pre-life conditions back in his lab at the University of California, Davis, to find that fatty molecules such as lipids could easily have been present in the conditions and with the chemicals available on the early Earth. Looking for further sources of these molecules, he hit upon looking for them in meteorites. He took a small sample of the Murchison rock (less than 1g), crushed it and took out some tiny fluorescent yellow particles that studded the grains. He knew that the particles were non-polar, though because he had such a small sample he could not identify the compound itself. When he stirred the non-polar chemical into a few millilitres of water, he saw tiny yellow droplets forming in the soup. 'Over a period of several minutes,' he wrote, 'these structures grew to form large numbers of membranous vesicles and long strands. The vesicles were clearly membranous and had open interior spaces in which Brownian motion of smaller particles could be observed. The largest vesicles were 50 micrometers in diameter.'

His work in 1985 was the first to show that raw ingredients from meteorites, which are commonly available throughout the solar system, could self-organise into membrane-like structures in water. Pohorille says the work was 'a great conceptual experiment that shows that you can form these kinds of structures very easily, even under conditions that could be comparable to those that may have existed on the early Earth'.

We now know more about how these membrane-like structures work. Biological cells on Earth are bounded by phospholipids, a long-chain amphiphilic molecule, which means that these molecules have one electrically charged end made from phosphate (let's call it the head) that is water-loving, while the other end (the tail) is an uncharged hydrocarbon chain that tries to avoid water. When the phospholipids end up in water, they spontaneously organise themselves into spheres made of two layers of molecules. If you made a cross-section of the sphere, you would see that the circumference of the circle is made from two molecules lined up tail to tail. The outside and inside of this 'bilayer' membrane is hydrophilic and in constant contact with water, whereas the inner part of the membrane, the hydrophobic part, is safely kept away from any water. (In this regard, they are not so different from Bangham's liposomes.)

That it is so straightforward to produce cell-like boundaries and membranes from common chemistry is important for the evolution of complexity. The reactions and chemistry of a life form have to happen in partial seclusion from its environment. 'Otherwise everything gets diluted in this huge primordial soup,' says Pohorille. 'The ability to form these structures where you can concentrate material inside and keep it inside is an enormous, very important, first step towards the formation of any kind of life or complex systems that could lead to life.' Water, specifically its polarity, was essential for forming the structures that became cells.

Compartmentalisation in cells allows useful molecules to concentrate in one place, more so than could be achieved if the metabolism that created them was just happening in an open sea, where everything could easily float away. It also protects the important molecules of an organism – everything from catalysts to genes to structural elements – from an outside world which might have intolerable pH or salinity or whatever else. Finally, cell membranes allow an organism to control what comes in and out of its intracellular environment.

Modern cells have taken the simple lipid membranes and run wild. Your complex cells are dense, viscous environments filled with sub-compartments – the nucleus, mitochondria, the Golgi apparatus, lysosomes and chloroplasts among others – as well as proteins, ribosomes and countless other biomolecules. And the water between it all is unlike anything you would recognise sitting in your glass or rolling around the oceans – it is more viscous and, some have argued, more like a gel or a liquid crystal than a liquid.

Understanding the nature of water inside cells is a relatively new science and we will need to pause our discussion of it here, or else it will take over the book. But it is a reflection of something that will come up time and again, that the most surprising things about water are sometimes so surprising simply because they have been right under (or in) your nose and you've been carrying that strangeness around with you this whole time.

Water's polarity allowed the cell membrane to exist. This is just one example of self-organisation of otherwise inanimate molecules. But water's polarity also plays another key role in the origins and mechanics of life – it ensures that complex biomolecules such as DNA and proteins are the correct shape to do their jobs.

Proteins are polymers made from long chains of amino acids, often hundreds at a time. The DNA inside a cell nucleus contains instructions for how to string the amino acids together for each protein and this information is communicated via RNA, a single-stranded version of DNA, to a cellular machine called a ribosome (itself a protein). Ribosomes find the right amino acid units in their surroundings and combine them (each combination reaction, by the way, involves locking together amino acid units by throwing out some water molecules) and continuously churn out long, straight protein molecules. But if you could look down a microscope into a cell, you would never see functioning proteins in this form, as straight chains. Instead they are tightly packed structures, almost spherical in shape. And, for a protein, its three-dimensional shape is everything, crucial to its ability to function.

The final stage of finishing off a protein after it emerges from a ribosome is for the straight chain to fold itself into its correct three-dimensional shape, something that will allow it to work as an enzyme or to create some structure in the body. Working out how that protein folding happened took scientists most of the first half of the twentieth century.

The story begins with Walter Kauzmann, who was born in Mount Vernon, New York, and grew up in nearby New Rochelle. At weekends his father took him to New York City, where the young Kauzmann developed a taste for music that would last him a lifetime, specifically symphonies and opera. He had a chemistry set at home and, at school, a gifted physics teacher who sparked his interest in science. After getting a PhD from Princeton University in 1940, he worked at Westinghouse and was then recruited to work on the top-secret Manhattan Project to build the first atomic bomb, in Los Alamos, New Mexico. When the Second World War was over, Kauzmann returned to Princeton where he was made a member of the chemistry faculty and gained a reputation for being widely read and for his informal approach with his students – one year he taught his first-year undergraduate chemistry students a class on paper-making, simply because he had become interested in it and had visited a paper mill to find out how it was done.

Biochemists at the time were starting to ask a lot of questions about proteins – not only what their molecular structures might be but how they remained stable and how they worked inside cells. Kauzmann had heard seminars from the various thinkers in the field during his PhD years at Princeton and decided to tackle those questions from a different angle, since his background was in physical chemistry.

Kauzmann and his research team knew that proteins were strings of amino acids and that living things used only twenty different amino acids (out of the hundreds that could possibly exist). A mid-size protein made of around 300 amino acid units could therefore be arranged in any of 20^{300} ways, which is a huge number, bigger than the number of atoms in the solar system. Natural selection had whittled down the proteins used in living things on Earth to a number in the several thousands and Kauzmann set about trying to find some sort of order within that, a way to explain how these complex molecules came together.

The Nobel Laureate Linus Pauling had pointed out, in the 1950s, the importance of hydrogen bonds as a way to keep the structural elements of proteins stable but his work said little or nothing about how the molecules were arranged the way they were. A few decades earlier, surface chemist and Nobel Laureate Irving Langmuir had introduced the possibility that the hydrophobic (water-hating) effect was at play on proteins but, though he was accomplished in his own field, it took the endorsement of the crystallographer J. D. Bernal, who studied proteins, to make other scientists notice.

As Kauzmann put it, the potential importance of hydrophobic effect in protein folding was 'in the air', and he extended the ideas with one of his own in 1959. Studying the thermodynamics of proteins in solutions, he proposed that charged amino acids (in other words, molecules that were polar) would tend to surround the outside of a three-dimensional protein globule, whereas the non-charged (non-polar) amino acids would congregate on the inside. He called this the 'hydrophobic bond'. His idea was published a year before John Kendrew and Max Perutz used X-ray crystallography to discover the first protein structures, of haemoglobin and myglobin. Kendrew and Perutz won a Nobel Prize for their work in 1962 and praised Kauzmann's prediction. 'The most striking feature common to all globin chains is the almost complete exclusion of polar residues from interior sites,' they wrote. 'This is a remarkable vindication of the predictions of Kauzmann in 1959.'

We now understand that some sections of the protein molecule are hydrophilic (they are polar or can engage in hydrogen bonding) while others are hydrophobic (perhaps because they have fatty or oily chemicals attached). As the long, floppy protein emerges into the cellular water from the ribosome that makes it, the molecule is nudged to fold into its correct three-dimensional shape because some of its parts want to be near water and some parts want to minimise contact. All of this structure gets built automatically, governed only by the natural physical forces that exist between the molecules that have been built with varying polarity along their length and the water in which they are bathed.

In tracing the discovery of how proteins fold, protein biochemist Charles Tanford wrote that the hydrophobic force had become the dominant force for explaining many life processes, including containment, adhesion and so on. 'This means that the entire nature of life as we know it', he concludes, 'is a slave to the hydrogen-bonded structure of liquid water.' So, water is fundamental to how cells form and to how proteins take their correct shape within those cells. The basic building blocks of life are intrinsically linked to water. But water's role in life doesn't stop there. It also, it turns out, plays a crucial role in how we access the energy necessary for life to function.

Cells need energy to do their work, whether they are muscle cells that need to contract, nerve cells firing electrical impulses, pancreatic cells making insulin or bone marrow cells producing blood cells. The energy these cells need starts from your food – every bite you eat is broken down into successively smaller bits as it makes its way to your cells. First you chew, then your stomach acids and the enzymes in your small and large intestines break apart the chemical bonds in your meal until the constituent nutrient molecules – carbohydrates, fats and proteins – can be absorbed into the blood.

Carbohydrates, such as glucose, are the body's main source of external energy but your cells cannot use this fuel directly to do its work. Instead, mini power plants inside your cells called mitochondria break apart the glucose molecules by reacting them with the oxygen you breathe. That process, known as aerobic respiration, releases energy which is then stored in a more useable form, a molecule called adenosine triphosphate (ATP). This is the energy currency of the body and you make vast amounts of it all day long – at rest you will produce the equivalent of half your body weight every day, rising to several times your body weight if you are working hard or exercising for part of it.

By the start of the 1960s, scientists knew a lot about

ATP and how it powered the body's cells. But how the energy from aerobic respiration actually ended up making ATP was a black box. Many biochemists had tried and failed to peer inside but it took a radical idea by an independent scientist based in Bodmin, Cornwall, to finally break it open.

Peter Mitchell had gained a PhD from Cambridge University in 1951 for research on how penicillin worked in the body. He spent a few years as a postgraduate at Cambridge and moved in 1955 to set up a research unit, the Chemical Biology Unit, in Edinburgh University's department of zoology. Though his work went well and he published some of his most important papers while there, Mitchell never really liked his Scottish environment. His colleague Murdoch Mitchison recalled that 'Peter liked to grumble about having to use a Land Rover to get through the snow and seemed to feel that it was a malignancy directed at him. He did not appreciate the local hills and their magnificent walks, nor did he have much sympathy for the Scots.' Mitchell was not naturally gregarious, preferring the company of a small group of colleagues over attending large meetings. In 1963 he fell severely ill with gastric ulcers and, unsure whether or not his research unit would be made permanent, he left Edinburgh University and withdrew from scientific research for the next two years. Over that period, he used this time to supervise the renovation of a mansion, Glynn House, in Bodmin. He farmed the nearby land and milked eight cows by hand every morning and evening for several months, something he claimed was excellent therapy for his ulcers. The independent research institute he founded in the mansion, Glynn Research Ltd, would be the base for the rest of his career.

In 1961, while still in Edinburgh, Mitchell took the challenge of trying to crack the mystery of how ATP was made by cells. His idea was radical and, at its core, showed that to power all of life on Earth you just need to store up protons (the bare nuclei of hydrogen atoms ripped from water molecules) on one side of a cell membrane and, when you need to use the energy, let the protons flow back to where they started. And the best and most efficient way possible to move protons around is in water itself.

The hawala system is a way of transferring money between people without going through banks. Developed in India (where it is called the hundi system) around the early Medieval period, it became widespread between traders in Europe and Asia before traditional banks took over.

It works like this: say you live in New Delhi and need to send Rs 1,000 to a friend who lives in Mumbai, as soon and as safely as possible. You give the money to a local hawala broker (known as a hawaladar), who extracts a small fee and then phones his hawaladar colleague in Mumbai. The hawaladar in Mumbai records the details of the transaction and information about the payee. Your friend in Mumbai visits this second hawaladar and, after some identity checks, is given Rs 1,000 in cash. The hawaladars settle up between themselves at a later date.

The whole process is informal, in that it doesn't go through state or commercial institutions, and relies on trust. Your friend in Mumbai could get your money within hours of you sending it, perhaps even minutes, much faster than any traditional bank. Today, hawala is still in use around South Asia and many parts of Africa, mainly an informal, parallel mechanism to banks, a way for migrant workers to send remittances back home. Hawala is so effective because users can quickly transfer cash around a network of brokers, without the hassle of physically moving any banknotes around.

Something similarly efficient is happening in your cells right now. The cellular hawala system does not transfer cash, though, but protons. More precisely, it transfers the energy and charge of protons from one part of a cell to another at impossible speeds, via networks of hydrogen-bonded water molecules. That movement of electrochemical energy ultimately powers everything that goes on inside you.

This nano-scale hawala system for protons is more formally called the Grotthuss mechanism, after Freiherr Christian J. Theodor von Grotthuss, a chemist born in Leipzig in 1785. The son of an aristocratic German family that had moved from Westphalia to the Baltic region around the end of the eighteenth century, Grotthuss carried out experiments on the decomposition of water into its constituent elements, using early types of electrical batteries, while in Naples and Rome in 1805. A year later, he published a landmark paper suggesting that water contained positive and negative 'corpuscles' (which he thought were H and O respectively) that disassociated when placed under an electrical field. (The familiar, and correct, H_2O formula was still unknown at the time.)

We already know that the hydrogen bonding between water molecules allows it to form vast three-dimensional networks of molecules. But this bonding also allows water molecules to form long chains, often called 'water wires', not only because of their one-dimensional, wire-like shapes, but also because cells use these chains to conduct electrical charge in the form of protons.

Protons have a special relationship with water, and not least because each water molecule contains two protons bonded to the oxygen atom. While other charged particles involved in cellular functions – sodium, potassium and calcium ions, for example – have to move themselves physically from one place to another in order to do their jobs, protons can pass their energy along a hydrogen-bonded water wire without moving themselves at all, thanks to the Grotthuss mechanism. A proton can become attached to one end of the water wire and, within a fraction of a second, each of the hydrogen bonds further along the length of the wire spin around in sequence so that a proton drops off the water molecule at the other end of the wire. The initial proton has not moved any further than the starting end of the wire but its charge and energy have been 'conducted' along the wire's length. Water wires moving protons around like this are astonishingly good at moving charge and energy around and, as Peter Mitchell found, are at the heart of the power systems in living things.

Without much experimental evidence to back it up, Mitchell proposed that the energy released during the metabolism of glucose was used by the mitochondria to pump protons, along water wires, from inside themselves and into the rest of the cell. This set up a proton gradient across the membrane, in other words a higher concentration of protons just outside the mitochondria, compared to their insides. It also set up a slight difference in electrical charge across their membranes.

Whenever you set up a difference like this in any natural system, it will tend to want to rebalance itself – pump water to the top of a hill and it will want to flow back down again as soon as it can (it will flow down the physical gradient, pulled by gravity). For the mitochondria, that rebalancing means that protons will flow back inside (they will flow down the proton gradient). Where there is a flow of charged particles, there is an electrical current and that can be used as a source of power, much in the same way a flow of negatively charged electrons down copper wires powers your computer, your lights and everything else in your home. Mitchell's idea was that the current produced by the flow of protons into the mitochondria could be used by machines within the cell to make ATP.

The idea was counterintuitive and seemed convoluted, to say the least, and it did not go down well among other scientists. Many of his contemporaries believed there should instead be some as-yet-undiscovered molecule that was an intermediate, which stored the energy from the breakdown of glucose before it ended up in ATP. Mitchell's hypothesis involved a paradigm shift, wrote the Cambridge University chemist Leslie Orgel, and his ideas 'seemed bizarre to most of his contemporaries'.

For almost twenty years, Mitchell's hypothesis remained controversial. Orgel recalls a meeting he had with Mitchell, at the latter's request, just after he had developed the ideas around proton currents as a way for living cells to derive their energy. 'I was too polite to express a view on the likelihood of Peter's mechanism being correct, but I remember thinking to myself that I would bet anything that ATP synthesis didn't work that way,' he wrote. 'Fortunately, no one took my bet.'

Mitchell's ideas eventually held up to experimental tests and, in 1978, he was awarded a Nobel Prize in chemistry for his work. 'Not since Darwin and Wallace has biology come up with an idea as counterintuitive as those of, say, Einstein, Heisenberg and Schrödinger,' wrote Orgel.

At his speech to the Nobel banquet in 1978, Mitchell acknowledged how the scientific community had to test new ideas to destruction wherever possible. 'Meanwhile, the originator of a theory may have a very lonely time, especially if his colleagues find his views of nature unfamiliar, and difficult to appreciate. The final outcome cannot be known, either to the originator of a new theory, or to his colleagues and critics, who are bent on falsifying it. Thus, the scientific innovator may feel all the more lonely and uncertain.'

In his final words to the assembled dignitaries, he gave thanks to his colleagues and 'especially to those who were formerly my strongest critics, without whose altruistic and generous impulses, I feel sure that I would not be at this banquet today'.

What we know now is that the ATP molecule is made

by an enzyme embedded within the mitochondrial membrane, powered by Mitchell's current of protons coming from outside the mitochondrion and into the inside.

For Mitchell, the proton gradient discovery was a by-product of work on mitochondrial cells in general. 'Mitchell knew protons were important, but he could hardly have guessed at just how important,' writes biologist Nick Lane. Mitchell showed that proton gradients have a basic role in powering cells. But he might not have guessed that his idea might be more fundamental to life than even that – his electrochemical flows of energy could have been the spark for the origin of life itself.

There is general agreement among scientists that our life, or something like it, emerged around 4 billion years ago, some time after our planet had gone through its tumultuous early years and the oceans were in place and the environment was relatively clement for chemical reactions to start ratcheting up towards the complexity we know today as biology. How the first spark for that ancestor came about is unknown but ideas fall into two groups.

The first set of ideas is that the earliest organic molecules, which later built life, were created in a primordial soup of some sort. In a letter to the botanist Joseph Hooker, Charles Darwin speculated about a 'warm little pond, with all sorts of ammonia and phosphoric salts, light, heat, electricity etc present'. In this theory, the beginning of life is a matter of the right chemistry.

In the middle of the twentieth century, two scientists at the University of Chicago put Darwin's idea to the test. Stanley Miller and Harold Urey filled a flask with boiling water, methane, hydrogen and ammonia, chemicals that were thought, at the time, to be abundant on the young Earth. The scientists passed electric charges through the mixture to simulate the lightning strikes or other sources of energy that might have been present at the time. The experiment worked – when Miller and Urey opened up their flask they found that the electricity had sparked the creation of amino acids, the building blocks of proteins. The flask also contained many other of life's key organic molecules such as sugars, fats and nucleic acids.

The alternative set of ideas for the initial phases of life starts deep underwater, on the sea floor near the cracks in the Earth's crust where our planet's internal heat spews out of hydrothermal vents. These areas of the ocean floor teem with life, with everything from bacteria to crabs that have evolved to take advantage of the scalding hot water and the rich supply of minerals that bubble out around the vents. The first hydrothermal vents discovered – in the early 1980s – and the most widely known are the so-called 'black smokers', which churn out acidic water and give out heat at hundreds of degrees Celsius. They support a wide variety of life today but these extremes might not have been the best conditions in which to begin the delicate early stages of life. Any naked organic molecules that formed would likely have broken apart at the temperatures found at black smokers.

Instead, geologist Mike Russell of NASA's Jet Propulsion Lab in California proposed an idea in 1989 that a gentler type of hydrothermal vent, which put out alkaline water at only 100°C or thereabouts, could be the site of the origin of life. Russell suggested that, in an ancient vent system like this, there would be a steady supply of hydrogen, carbon dioxide and minerals to act as catalysts in the necessary chemical reactions. The whole thing could go on in the pores inside the rocks (which would mimic the functions of a cell membrane before such things had evolved and prevent the products of one reaction from becoming diluted or drifting away in the open ocean). For a power source to keep the pre-biological reactions going, Russell proposed natural proton gradients.

Four billion years ago, the Earth's oceans were mildly acidic because there was much more carbon dioxide in the atmosphere and some of this would have dissolved into the oceans to make carbonic acid. The acidity of a solution is just a measure of the concentration of the protons present and the acidity of ocean water in the young Earth's oceans would have been several orders of magnitude (in other words several pH units) higher than the alkaline water coming out of the hydrothermal vents. What makes this idea more compelling is that, in Russell's scenario, the difference in proton concentration between the acidic ocean and the alkaline entrance to the vent would have created an electrical gradient that could provide the same level of energy for cellular processes as that in modern biological cells. That proton gradient, Russell says, could have powered the start of life. One reaction could have led to another and so on, building ever more complex molecules.

For a long time, however, there was a big missing piece to Russell's theory. When he suggested it, no one had ever seen a gentle alkaline vent. It took more than a decade from Russell's prediction but marine biologists finally discovered an alkaline vent field in the mid-Atlantic in 2000. They called it the Lost City – dozens of mineral chimneys, some 60m high, dotted the sea bed, created when water had been pulled into the cracks in the floor, heated up and then pumped back out into the cold ocean. As that had happened, the calcium carbonate dissolved within the water had precipitated out and crystallised into the magnificent structures that contain the types of pore-filled rocks Russell predicted for his pre-biotic world. The water coming out of the cracks on the sea floor can get up to 90°C and the vent field supports a rich diversity of snails and amphipods. Inside the vents themselves are thick mats of archaea that live on methane and hydrogen.

There are still plenty of unanswered questions in Russell's

idea. If proton gradients were the power source, how did the precursors of biological cells actually use that energy? In modern cells the energy flow of protons powers the production of ATP in mitochondria, but the porous rocks around alkaline vents would not have had such sophisticated mechanisms. Still, Russell's is a bold, attractive idea that relies on nothing but the basic laws of thermodynamics to set up an energy flow that, somehow, some molecules managed to harness and with it get a foothold into creating the complex life we see today.

Most impressive, perhaps, about the credibility of this idea is that it shows the repeated power of Peter Mitchell's lone idea that protons, moving through water networks, are the spark that makes life work. From the core of metabolism to the best candidate for the origin of life, Mitchell's revelation about the role of water, and its constituent parts, in life is more profound than he might have known.

SIX

Water Footprint

L eonardo da Vinci was already a famous painter, architect, sculptor, anatomist and musician when he made his way to start a new job in Imola, northern Italy, in the summer of 1502. He had been summoned by Cesare Borgia, the ruthless and power-hungry Duke of Valentinois. Cesare had recently taken control of large parts of northern Italy and was looking west for further conquest, to the stubbornly independent city-state of Pisa. He needed an advisor in military engineering, and Leonardo, who was also an accomplished inventor, needed a patron.

What the gentle Leonardo made of his bloodthirsty employer is not recorded. Cesare came from a family that, according to historian Roger Masters, amassed power and wealth 'by the traditional Italian means of family and cheating'. He had been groomed for a life in the church, becoming the Bishop of Pamplona by the age of fifteen and Cardinal of Valencia three years later, when his father ascended to the papacy and became Alexander VI. Even as he rose through those hallowed ranks, however, the young man craved a different kind of power. His older brother, Giovanni, had been given command of the papal armies, entrusted with their father's military ambitions and the might of the Borgia family. When Giovanni was assassinated under mysterious circumstances in 1497 – some suggest his younger brother was responsible – Cesare was already poised to take over. He resigned as a cardinal the following year and took command of his father's armies, using them to seize control of a chain of cities across northern Italy.

As an engineer at Cesare's court, Leonardo worked on designs for everything from strengthening castle defences to blueprints for temporary bridges that a rampaging army might build in the field, in order to ford a river on campaign. His most ambitious idea during this period, though, was spurred on by another visitor to Cesare's court.

Niccolò Machiavelli, chancellor and secretary of the nearby Florentine Republic, had been sent to Imola to find a way to work with Cesare against a common enemy, the Pisans. He was also there as a spy, watching for anything that might indicate that his ambitious, bloodthirsty young host had started making plans for an incursion south, perhaps to Florence. Machiavelli was already a rising political star when he met Leonardo. He wanted to find a way to bring Pisa to its knees without the mess and cost of battle. He turned to Leonardo for inspiration and the duo, two of the greatest minds of the Renaissance, subsequently worked out a plan that was as ambitious as it was ingenious.

Machiavelli and Leonardo decided they would divert the River Arno, which flowed through Florence before going on to Pisa, and so deprive the Pisans of their vital connection to the Mediterranean Sea. Eventually, without a secure way to bring in food and supplies, a hungry, thirsty Pisa would be no match for the invading Florentine or Borgia armies. For Machiavelli, there was the added bonus of a direct link to the sea that would be controlled by Florence: a potential source of increased wealth and prosperity for his beloved republic.

Leonardo was excited about the idea for other reasons, too. He had seen the Arno burst its banks many times, and watched as rising waters swept away people, animals and buildings. His complex system of canals and reservoirs could help to control the floods, directing them, instead, to irrigate Tuscan farms. Here was a chance for him to gain some control over a substance that he had spent decades studying, drawing and trying to fathom. For Leonardo, water was the 'vehicle of nature' (*'vetturale di natura'*), the driving force behind all natural things. He was obsessed with it.

Water, he reasoned, was the fluid that transported nutrients around the Earth, feeding plants and fields, just as blood, according to Galen, nourished the organs of the human body. He knew that water was the medium of life: that life must have started in water, and could only exist in its embrace. He inferred the presence of a hydrological cycle that moved water around the world, from seas to mountains to rivers and back again. Before the scientific method was even a whisper in the thoughts of Isaac Newton or Francis Bacon, Leonardo was dutifully measuring the physical characteristics of water as it moved around in its various forms – as a vapour, as drops on leaves, in streams, as ice and snow.

Leonardo was the first to record the fact that water flows more slowly at the bed of a river and its banks than it does at its centre, because of the relative increase in friction at those places. And he calculated that the overall speed of a river at a given point is inversely proportional to the area of its cross-section at the same point. In other words, when a river narrows, it speeds up.

Being Leonardo, of course, he also tried to apprehend water visually. He filled sketches with maelstroms of lines, attempting to portray the vortices and eddies as water moved around objects, or fell from a height into a still pool. He designed machines that could move water around in bulk.

But Leonardo's interest in water wasn't only practical: while other engineers of his time were content to understand water enough to build dams and irrigation channels, Leonardo looked for the mathematical underpinning, and created a nascent version of the science of fluid dynamics, puzzling over the ways friction and viscosity affected water's movements. More than three centuries before Hermann von Helmholtz formalised his theorems on the mechanics of a vortex, Leonardo had already noticed that the water at the centre of a vortex moved faster than the water at the edge. This, he pointed out, was the opposite of a wheel, the speed of which slows the closer you are to the centre.

In his drawings, the individual currents of a body of water are imagined as filaments or hairs, moving in many different directions simultaneously. These sketches portray both change and continuity. Swirling, moving water was not just a passing curiosity for Leonardo. He clearly feared it, too. His drawings and his writing are full of an anxiety about what this tumultuous, untamed substance might do next. He wanted to control it. The plan he hatched in 1502 was more than a technical challenge for a master engineer – in Leonardo's imagination, to tame water was to tame the natural world, to extend the influence of his ordered mind over the powerful and unpredictable forces he saw outside it.

Of course, it never happened. For all the combined brilliance of the two men at its helm, the plan to divert the Arno failed. The canals built for the project were too shallow. There were storms too, swelling the Arno, and the raging river burst through the dams that had been built to hold it back while work progressed. Dozens of labourers died. Though Machiavelli and Leonardo lay low for some time after the debacle, both ultimately moved on to greater things. Machiavelli used his forced temporary retirement from public life to write *The Prince*. Leonardo used his to begin work on the *Mona Lisa*.

But Leonardo kept on with his study until his death in 1519. His Codex Leicester is as much a treatise on water as it is the record of a pre-scientist trying to move from a spiritual to a scientific understanding of nature.

Several years after the Arno project, Leonardo summarised his feelings in a passage that itself flows from attribute to attribute, as though his words are both describing and enacting the mutability of water:

Water is sometimes sharp and sometimes strong, sometimes acid and sometimes bitter, sometimes sweet and sometimes thick or thin, sometimes it is seen bringing hurt or pestilence, sometimes health-giving, sometimes poisonous. It suffers change into as many natures as are the different places through which it passes. And as the mirror changes with the colour of its subject, so it alters with the nature of the place, becoming noisome, laxative, astringent, sulfurous, salty, incarnadined, mournful, raging, angry, red, yellow, green, black, blue, greasy, fat or slim. Sometimes it starts a conflagration, sometimes it extinguishes one; is warm and is cold, carries away or sets down, hollows out or builds up, tears or establishes, fills or empties, raises itself or burrows down, speeds or is still; is the cause at times of life or death, or increase or privation, nourishes at times and at others does the contrary; at times has a tang, at times is without savour, sometimes submerging the valleys with great floods. In time and with water, everything changes.

Our control of water today is so precise that we barely think about it. We turn on a tap and it gurgles into life to produce a flow of clean, fresh water. We don't see the intricate machinery that brings all of that into our homes and offices

– the maze of underground pipes and tunnels that honey-comb the ground beneath our feet, the sewerage and water-processing facilities around the edges of the towns and cities in which we live. Nor are we likely to think much about the thousands of years it took to conceive, design and build this mass movement of water. An artificial network of streams and rivers, bounded by metal and cement, and designed for the singular purpose of keeping people alive.

Around 2,000 years ago, the Greek geographer Pausanias travelled the ancient city-states of his region and, in his writings, concluded that no city could call itself a city without an ornamental fountain at its centre. 'Fountains illustrated then – as they have done throughout the ages – an ideological and cultural notion of the triumph of civilisation over nature: water, the giver and taker of life, in the fountain appears at the control of human beings,' wrote geographer Terje Tvedt in an essay for his multi-volume series *A History of Water*. 'The fountain also symbolises a more mundane and direct material fact – no city and no country has been able to exist or develop without subjugating water in one form or another to the demands of human society.'

Fountains are an artistic expression of the human control of water, a universal and central project in the development of civilisation. Every major city in human history has been on or next to a river and it is unthinkable to have a significant concentrated human population without a replenishing source of water. At all stages of our development as a species, we have had to find ways to reliably find water or bring it to where we live. The management of water resources has changed landscapes and shaped society itself because those who control water, control people. Water is a source of political power and access to it created hierarchies and class divisions that still exist today. Artificial irrigation increased how much we could grow, how many people could survive off the land and enabled the division of labour crucial to the growth of the first cities. Flowing water turned water mills and was the main source of power for the thousand years before the Industrial Revolution. 'Dams have stored, regulated and raised water. Watersheds have been reworked and linked. Rivers have been forced between levees and dykes, canalised, straightened and cemented,' wrote Tvedt. 'Water has been diverted from areas of water surplus to areas of water deficit. Lakes have been lowered and wetlands drained and the artificial river is definitely not a modern invention.'

Our species has wandered the Earth for a few hundred thousand years and, for most of that time, we existed as tribes of hunter-gatherers. Our foraging ancestors went to where the food was and moved in bands of a few dozens of people, up to fifty on rare occasions. In the arid conditions of what is now northern Africa, they would have hunted small game and used stones to grind and pound nuts and cereals to extract food. They went where the water pooled into lakes or flowed in rivers – not only to drink it themselves but because that was where the plants and animals they would rely on also lived. Weather was critical to the survival of these early human populations, which lived and died on the seasonal rains. If there was a drought, there would be trouble and so people never strayed far from natural sources of water wherever they could. Options to store and carry water were limited – people might have fashioned containers from bamboo or wood to take water on hunting trips or, if the birds were present, the shells of ostrich eggs. That later became easier with the invention of pottery.

The nature of society in places such as the Nile valley in pre-history was, in essence, tribes that were expanded families, each with a chief and each one forming a particular lineage or ethnic group. The earliest hints of the structures that would become the governments and states we recognise today started when people developed farming, around

12,000 years ago, and the earliest civilisations – large group-ings of people organised by social and political heirarchies – grew up around Mesopotamia and Egypt around 6,000 years after that. The centres of all these empires revolved around access to water.

The move from nomadic tribes of hunters and gatherers to permanent settlements near rivers and lakes was an important step in our use of water. Now humans started to move water from its natural settings in ponds, lakes and rivers and into crop fields and other places where we had decided to set up home. The systems and the planning became more complex as settlements grew and eventually merged to become cities and then states and we needed more water brought to the rapidly growing urban areas of the world. Moving water in this way, using canals and drains, mitigated against the often patchy rainfall in the arid regions. These man-made channels and dams were the first attempts at irrigation, to manage water to feed growing bands of people who stayed in one place and no longer wanted to be at the mercy of the seasons.

These earliest methods of agriculture did not insulate groups of people from famine completely. Floods or droughts could still ruin their crops, making people go hungry or killing them outright. Neighbouring settlements of people began to make deals with each other to buffer against famine – sharing grains when they could and helping each other to build canals or repair damage when waters overflowed them.

These groupings of settlements grew in size and made more agreements until, by about 5,200 years ago, entire regions containing many villages were united and ruled by kings. The leaders protected their proto-states from attack and, crucially, managed the collective water resources. That became increasingly important as these regional societies grew and the people within them became stratified. In the earlier, smaller societies, people grew and harvested their own crops; in the larger, regional groupings, new groups and classes of people emerged who had specific roles that did not include time to grow their own food – priests, artisans, nobility and soldiers. Farmers had to work not only to feed themselves and their families, but also to produce food for these elevated elites in their state. That required more water and more technology – waterworks, dams and more sophisticated irrigation – to grow more from the land and ensure that the food was guaranteed in spite of the fluctuations in annual seasons.

People thought up plenty of innovative ways to bring water to cities and areas of the land that would otherwise have remained dry. By 2500BC, for example, the Persians had come up with an ingenious method of artificial irrigation known as the 'qanat' – a system of underground tunnels and wells designed to collect and transport water over tens of kilometres, particularly designed for arid areas. By reliably transporting water from mountain springs to farmlands at the edge of deserts, people could grow food on ever-larger areas of land. The very earliest remains of qanats have been found in Egypt, introduced there by its Persian neighbours, and the technology spread far over the following millennia to the Arab states, Pakistan, Afghanistan and China. Qanats eventually hopped across the Atlantic Ocean in the middle of the second millennium AD to Mexico, Peru and Chile after these water systems were introduced to Spain by Muslim visitors.

The hydraulic engineers in the ancient Roman Empire were masters of water. At the height of the empire's power, the area around Rome itself had a population of approximately 500,000 people, about ten times that of earlier urban centres. Alexandria in Egypt had around 400,000 people at its peak.

Both cities developed sophisticated mechanisms to move and store water to keep their inhabitants happy. Alexandria sat above a network of hundreds of water-storage tanks fed by a branch of the Nile. In Rome, a 10-mile under-ground tunnel known as the Aqua Appia provided the city with 16 million gallons of water per day. Soon that was not enough to supply the bathing and drinking needs of the city's citizens and so, by 140BC, the Aqua Mercia was built on elevated arches to transport 100 million gallons of water to Rome from the springs of Subiaco.

Sextus Julius Frontinus, Rome's water commissioner towards the end of the first century AD, used to complain about how the ornamental Egyptian pyramids and Greek temples got all the public's attention whereas the vital system of aqueducts that kept Rome watered were ignored. The Pantheon and the Colosseum may have brought Rome fame for their architecture and entertainment respectively, says Tvedt, but the city owed its existence to the water running beneath it. Aqueducts made Rome possible.

Across the Mediterranean, the Nile and its associated valley was (and still is) the lifeline of the people who lived around it. The great river flows through a country where rainfall is sporadic, sometimes with many years between any rainfall events in the driest parts of the desert. Cairo has an average rainfall (in millimetres) in the low twenties every year. Yet, this country was one of the main suppliers of food to the Romans and was even, a century ago, a cotton farm for the British Empire. 'The explanation for these "miracles" has always been the Nile and its annual flow over the coun-try's borders, under a cloudless sky, like an umbilical cord surrounded by sand,' writes Tvedt. 'The character of Egypt's water landscape has always made Nile control a top priority for its inhabitants and rulers. This is the case now, as it was about 5,000 years ago, when the "Scorpion King" was portrayed commanding the Nile to flow out over the fields in a canal dug by men, and Menes, the first Pharaoh, dug a new course for the Nile to protect his capital.'

Managing the Nile and its annual floods was at the

centre of Egyptian society and allowed it to rise to the prominence it did in the ancient world. Engineers and scholars developed ways to predict the river's flow, farmers worked out the best way to use the fertile silt-covered land created by the floods, and administrators and politicians developed ways to apportion and tax it all. This river being the main, and sometimes only, source of water for Egyptian societies since then, it still looms large in their recent history. Since the construction of the Aswan High Dam in 1971, the great river has been completely subjugated to human control. The dam creates a 500km-long lake, called Lake Nasser, near its southern end that reaches into Sudan. It not only provides Egyptians with the ability to control flooding of the Nile valley, giving them a hugely increased area for growing crops, but it also provides electricity thanks to the huge built-in hydropower plant. The Egyptians became masters of the Nile with this technology but, as a result, they solidified their dependence on it.

Elsewhere as cities grew in the first millennia AD, people needed water for washing and crops as well as, increasingly, for mechanisation and industry. 'In Europe, the development of Medieval towns linked to trade and crafts in a climate of competition and warfare not only made it necessary to secure water for city dwellers, but also made use of water for defence, mills, tanners and paper makers,' writes the archaeologist Fekri Hassan. 'Sewage, sanitation and water pollution became issues of concern, and had a major role in transforming water management methods.'

By the fifteenth century, affluent cities around the civilised world had wide tree-lined avenues and fountains, the 'symbols of the triumph of the city over its water problems and of its prosperity and affluence', according to Hassan. 'Fountains were, in fact, the new temples to water gods.' Pausanias would no doubt have been pleased.

Around this time, the land that would become Italy was

in the early stages of the Renaissance. This flowering of ideas included the rediscovery of the texts of antiquity, the work of philosophers such as Plato and Aristotle, pagans who had been largely sidelined by the religious leaders until then. Those texts and ideas reconnected human beings with nature and both Leonardo da Vinci and Niccolò Machiavelli were steeped in and fascinated by this thinking, albeit from different angles. Leonardo wanted to understand and control nature; Machiavelli was interested in how that knowledge could be used for political power. Their plan to divert the Arno meshed those ideas together.

The Arno is a windy river that flows from Florence to the sea, not easily navigable by large ships that might be used for trade. For a long time, the Florentines and others had been annoyed with the Pisans levying taxes on anything they wanted to send downriver and this had led to several wars. But there was more to Machiavelli's interest in Leonardo's idea to straighten the Arno than just subjugating Pisa. He saw it as the way to make Florence a seaport onto the New World.

At the end of the fifteenth century, Christopher Columbus and Amerigo Vespucci, among others, were opening up the Americas to Europe. Their mode of travel was, of course, by sea, and the places that would reap the most rewards from the new lands back in Europe would be places that had ports to deal with the copious crops, gold and other bounties that were flowing back from the land they plundered. Florence was already a commercial and banking centre at the time and overland travel from there to the rest of the Italian peninsula was easy and established. Travel by road is fine for letters or small quantities of goods, but not so great for heavy trade. On a road, a convoy of carts pulled by hungry horses is not only slow, but it might get attacked by marauders at any point. Sending goods downstream on a river was, in comparison, relatively efficient and safe. Developing direct water-based routes to the new worlds became a priority for the Florentines and, for that, they needed a port.

'Once Leonardo and Machiavelli had met, if we think of it in modern terms, it was one of the first attempts of a great scientist to put science at the service of a powerful, secular government,' says historian Roger Masters, who wrote about the Arno project in his book *Fortune Is a River*. Their combined efforts preceded what Francis Bacon would, a century later, call the conquest of nature 'for the relief of man's estate'.

Their project was ahead of its time and failed because they underestimated the strength of the water's flow in the Arno. But the idea of subjugation of water was not impossible and, crucially, the two realised that, in water, there were multiple dimensions for the advancement of a city and its people. People knew, of course, that cutting off the water flow of a city would destroy that city. But they also knew that opening up the access of an inland city to a watercourse was important for its commerce and that it would improve the economics of the city. Further afield, for explorers, the discovery that there was a huge body of water beyond the Mediterranean that nobody knew was even there between Europe and another continent meant that water was the only way of getting to new places. Bodies of salt water were more important than anybody had ever known.

Water was also an important source of physical power to run machines using water wheels and other similar devices. Those mechanical concepts in particular were made tangible in the machines designed by Leonardo. He wanted to control the natural forces of gravity and water to make human life easier and to grow human wealth and power. One way to do that was to create canals to help water avoid obstacles.

The use of water to create power would come to its

greatest height in the Industrial Revolution with the development of the steam engine by British engineers James Watt and Thomas Newcomen. They and their contemporaries used liquid water's anomalous ability to hold on to a lot of heat and then expand extraordinarily when boiled into steam, first to construct a new physical world that dramatically increased the human ability to build and do work and, around the same time, create the new science of thermodynamics.

Even a simple timeline of the rise of civilisation would be beyond the scope of this book. But we can note something interesting about the increased use and control of water – that progress can be measured in how many people there are on Earth. In 6000BC, there were a million people on Earth; by 1900 that had risen to a billion. Today, there are more than 7 billion. The amount of water on the Earth, however, has been the same over all that time and, accordingly, the control of that water has become ever more important.

With the increase in the number of people, there have been rising strains on the supply of water in the past century. We have dug more canals and reservoirs in order to increase the available agricultural land (it doubled from 1900 to 1950 and again from 1950 to 1990). More than half of the world's people now live in cities, which require associated water infrastructure and control. And we have become increasingly industrialised: we need more water to produce our goods, clean and produce electricity. Water not only flows around us but it flows through us and everything we touch, taste and see.

Water is a component of everything we eat, the clothes we wear and the products we buy. Modern life functions because of the hidden (or embodied) water in our goods. It may not surprise you to know that it takes around 200 litres of water for a five-minute shower, around 8 litres to flush a toilet and a similar amount to brush your teeth. You can see this water as you use it and, if you are conscious of saving the resource, might even do something to stop overusing it.

But most of the water we use is hidden. It takes nearly 200 litres of water to grow the coffee beans in one cup and the same again for every cup of coffee you have during the day. It took 15,000 litres of water to grow the kilogram of beef you cooked for your family's dinner; the whole day's food took more like 35,000 litres of water. Even a small soy burger requires 160 litres of water to grow the plants from which it came (more economical than a 1,000-litre beef burger). It takes 100 litres of water to make the two slices of bread for a sandwich, 65 litres to make the cheese filling and around 250 litres of water for the milk in each of your lattes. When you get home at night and wash the dishes after dinner, you'll use around 75 litres of water in the sink.

This concept of the 'water footprint' was introduced in 2003 by Arjen Hoekstra, a professor in water management at the University of Twente in the Netherlands, as a way to link together human consumption with the amount of fresh water being extracted from the Earth. The footprint of a product – also known as its virtual water content – is a sum of the water use of all the processes needed to make it. For a loaf of bread that means growing the wheat, milling it, producing packaging, transporting it and so on.

You can calculate the water footprint of almost anything – a sheet of paper has a water footprint of 10 litres, a microchip is produced using 32 litres of water. Your clothes also come with a water price tag. A kilogram of cotton (a shirt and some jeans, say) will have required 10,000 litres of water to make, on average, though this figure varies wildly according to where you sourced it (cotton from India can 'cost' up to 22,500 litres per kilo).

And don't forget the beer you had as you sunk into your couch to take all this in. A typical pint would have used around 150 litres of water to make. And perhaps you're reading this on an iPad, or a Kindle. These devices collectively require millions of litres of water to wash, prepare and make.

The vast majority – around 70 per cent – of the world's use of fresh water is used in agriculture, mainly to grow wheat, maize and rice. That is not only for our own food but also to feed the growing populations of livestock in the world. A fifth or so of the annual water use goes to industry and the rest is used at homes and offices. The numbers shift around slightly depending on how developed a country is – lower and middle income countries use a bigger proportion of water for agriculture and less for industry, for example. And our use of fresh water is on the up: according to the United Nations, water withdrawals have tripled over the past fifty years as populations have grown. They are predicted to increase by 50 per cent by 2025 in developing countries and 18 per cent in developed countries.

There are some 33 million cubic kilometres of freshwater on the Earth in total. We each drink around a cubic metre of water every year and use 100 times that amount for cleaning and washing. The food we eat every year takes up another 1,000 cubic metres to grow. The Intergovernmental Panel on Climate Change (IPCC) and United Nations warn of a coming water apocalypse – there will be skirmishes between countries for decent access to water, they warn, and billions of people will lack access to clean water by the middle of this century.

Right now, more than 1 billion people in developing nations already find themselves in this position – without easy access to safe drinking water – and more than 2 billion lack proper sanitation. Increasing prosperity in places such as China and India add more pressure as the population (rightfully) demands better, more protein-rich, diets. These countries will also demand more water use for their energy sectors. Their consumption could easily get to the level of intensity seen in the United States, where 500 billion litres of fresh water are used every day to cool electric power plants. The country uses the same again on irrigation.

Around the world, energy demand is due to increase by 57 per cent by 2030 and the demands on water for agriculture could easily double by then. A report in the journal *Nature* pointed out that, by 2050, providing food for the world's population will require around 12,000 cubic kilometres of water, equivalent to the volume of Lake Superior, every year. Already the drop in available water shows: China's Yellow River doesn't always reach the ocean, and Lake Mead in the American south west could be dry by 2021.

In 1995, the UN predicted that half a billion people would be living in water-scarce or water-stressed areas by 2050; in 2005, the organisation revised that prediction to 4 billion people.

Climate models show that, as global temperatures increase towards the end of this century, extreme droughts will become common across large parts of the world, leading to the death of agriculture in those regions. Areas at risk will increase from 1 per cent of present-day land area to 30 per cent by 2100. In South America, the number of people living in water-stressed environments (in other words, with supplies of less than 1,000 cubic metres available per person per year) in the absence of climate change was estimated at 22.2 million in 1995. This number is estimated to increase by between 12 and 81 million in the 2020s and to between 79 and 178 million in the 2050s.

In 1940, the Chacaltaya Glacier in Bolivia covered an area of 0.22 square kilometres of surface ice; by 2005 this had reduced to less than 0.01 square kilometres. Over the period 1992 to 2005, the glacier suffered a loss of 90 per cent of its surface area, and 97 per cent of its volume of ice.

In Africa by 2025, nine countries in the eastern and

southern parts of the continent will only have 1,000 cubic metres of water available per person per year. According to an estimate in an IPCC report on water and climate change, published in 2008, the proportion of the African population at risk of water stress and scarcity will increase from 47 per cent in 2000 to 65 per cent in 2025. This scarcity will generate conflicts over water.

Water availability in India is projected to decline from about 1,820 cubic metres per year per person in 2001 to 1,140 cubic metres per year in 2050, as a result of population growth alone. One option available to the growing population is to dig more wells – but dig too deep or too many and the freshwater will become contaminated with sea water or arsenic. The groundwater is already becoming too poor to use in some parts of the country that depend on agriculture, leading to skirmishes between neighbouring villages or states over access to rivers or lakes. Thousands of farmers have committed suicide in the past decades as poverty and debt drive them to despair.

All over the world, we are overexploiting ancient aquifers. Our global groundwater footprint, according to scientists, is 3.5 times the actual area of aquifers around the world.

Once we might have seen water as a free, limitless resource that falls from the skies and fills seas and rivers ready for our exploitation, without much thought. That free use will not last forever.

Part III

CRYOSPHERE

SEVEN

The Ice

Antarctica officially starts at 60 degrees south. It's an arbitrary line, in truth, because the sea there seems much the same as the sea at 59, 58 or 57 degrees south – rough, roiling and deep blue. There are subtle changes in sea temperature, for sure, as you head south but these aren't evident just by looking. Out on the deck of the *Akademik Shokalskiy*, the air temperature had been dropping consistently, from 15°C at the start of the expedition to 13°C two days later, 8°C on our brief pause near Macquarie Island, 4°C the day before we crossed the Antarctic circle and, at 60 degrees south, 3°C.

If the winds were not blowing (yes, sometimes they did take a rest) and the Sun was high in the sky, it was possible to stand around outside with just a light jacket, perhaps even a shirt. On those days the near-zero air was refreshing on the skin and in the nostrils, cleaning away the seeming sluggishness that followed us around inside the ship. On the best days like this, being out on the observation deck, where the metal had been warmed by a few hours of direct sunshine, it felt like a still day late in the spring. All that was missing was the sound of birdsong and idle chatter, the smell of barbecues and the rustle of trees. Though the last of these was admirably replaced by the insistent splosh of water against the hull of our ship.

By this point in our journey, the Sun was doing an ellipse around the sky, rising and setting through the day but never actually disappearing. It was always light and had been for several days. Around 'sunset' – usually at 1 or 2 a.m. – the Sun would dip briefly to the horizon and flirt with the clouds there, turning the sky into a rich array of oranges, golds, reds and apricots. The colours were unlike anything I had seen anywhere else, a collection and movement of light that rendered any New Year's Eve, 4 July celebration or bonfire night firework show into a stick drawing in the presence of a gallery of old masters. Our Antarctic show was pure light, with the soundlessness only intensifying what we were seeing. And all of that was given an extra kick by the scale of our surroundings. We had clear views in all directions, not one thing between us and the distant horizon on all sides, and the sunset show lit up our entire hemisphere.

One evening, clouds peppered the sky, and the light glittered and twisted around them and bounced off the water in every direction. Everything here moved all day long and, at sunset, an evanescent layer of shadow and colour danced and shimmered across the surface of that world.

The Sun came up fast every night, though, after its dip towards the sea – by 3 a.m. the light show was always over and we were back to the starkness of daytime.

The morning after we crossed into the Antarctic circle, I woke up to the sound of bangs and explosions coming from outside the ship. There were ominous judders, creaks and rumbles as if the ship's hull was sliding, no, scraping against something outside. Most salient, though I only noticed it well after I had woken up, was the fact that we had stopped rolling. I ran over to the porthole in my cabin, pulled up the heavy iron shutter and wiped the condensation off the thick glass.

The sky was a grey wash. The sea was flat, almost unmoving, and the darkest I'd ever seen – it looked like a black mirrored glass. And the whole scene, from our ship to the edge of the field of view, was dotted with ice. At 63 degrees south, it finally felt like we were in Antarctica.

An almost-infinite number of ice floes were floating on the ocean around us, their sizes ranging from small cars to tennis courts and beyond. All of them mostly flat, a metre, perhaps two, above the surface (though many more beneath). They moved idly in the distance but our ice-strengthened hull was making quick work of carving through the ones in our path, cutting and churning them, turning them over to reveal hollows and marks where they had been eaten away by the sea water. Our ship seemed to be the only thing creating movement in this vast, still landscape. Every raft that smashed and split against the hull made the ship shake in return. The pieces of ice swirled around us as we made our path through it.

Out on deck, the air temperature was sub-zero and the ice had come onto our ship as well as surrounding it. All the metal surfaces glistened with a light carapace of frost and walking around required some careful thought – everything was potentially slippery and, anyway, it wasn't too easy to grab hold of anything too tightly while wearing the thick gloves needed to keep your hands from turning into icicles in these conditions.

The wind was light and dropped the effective temperature of the air by several degrees every time it gusted across the bow. Thanks to multiple layers of wool, insulation and various temperature-controlling technical fabrics, my torso, arms and legs all felt fine. It was a different story for the exposed parts, though, a gentle reminder of what the multiple layers were protecting me from. It took only a few minutes of standing on the bow for my nose to lose all feeling. Taking off my gloves for more than a few minutes to use a camera or write some notes quickly left my fingers red raw and stiff. This was just the beginning of the cold, I thought; we hadn't even reached the continent yet.

The surroundings helped us all take our minds off the edges of discomfort, however. We leaned over the edge of the deck and tried to get as close as we could to the ice – it looked even more dazzling from a few metres away. These pieces of sea ice, frozen bits of ocean, were the remnants of the previous Antarctic winter, a time during which the seas around the polar lands freeze, doubling the effective size of the continent. Every summer, as the temperatures rise, the frozen sea begins to melt and the ice breaks up into pieces. Moved by winds and water currents, these bits of sea ice float around and collect in huge patches around the polar seas. If they get far enough away from the continent, they will eventually melt into the Southern Ocean.

Where we first saw the sea ice, about 160km from the Antarctic coastline, there was plenty of open water between the floes, so the captain of the ship was able to steer a relatively straight course, pushing some ice floes out of the way, splintering through others. Closer to the continent we were told that the sea ice, pummeled by the fierce winds coming off the polar plateau, would collect into larger pieces. This pack ice not only promised to be thicker and denser in structure but also more comprehensive in its coverage of the sea surface, making it altogether more tricky to navigate through. The bounding progress of the ship through the Southern Ocean, and also through the ice field we had just encountered, would turn into a laborious grind in the approach to Antarctica as the captain looked for leads through the ice.

But that was all still to come. At that moment, 63 degrees

south, we not only saw the first ice but also some of Antarctica's native wildlife: Adélie penguins. I saw my first three on an ice floe ahead and to port of the ship, seemingly intrigued by our vessel for only a moment or two as we approached but resolutely bored by the time we were moving alongside them. The first penguins caused a commotion on deck as everyone scurried to find the best place from which to see them and capture the moment on two dozen cameras. We needn't have worried – after the first group of three, we saw new ones every few minutes on ice floes near and far. Flocks of them leaped along in the water next to the ship, others just sat or stood on the ice floes, unfazed, staring into the distance, occasionally plopping into the water.

As if the penguins hadn't caused enough squeals of excitement, at about midday when almost all the expedition members were out on the bow of the ship, we spotted an iceberg looming in the distance. My heart stopped.

You see icebergs in books, on TV and on posters. You know what they are, how big they are and that what you see is only a tiny fraction of its mass – below the water is an object of a size that defies any human imagination. You know all this. And yet, on seeing my first iceberg with my own eyes, I was momentarily incapable of comprehending what I saw.

To the untrained eye, it was unclear what we were looking at, at first. In the fog, it was hard to see edges or get a sense of the dimensions of this density of white in the distance. The captain steered the ship towards the berg and, like a developing photograph, it resolved into a sharp focus as we came up to a gentle cruise alongside it.

At least 50m high, hundreds of metres long, solid, white. Pure, magnificent white. As we got closer, I saw layers in the face of the cliffs and holes and caves across the surfaces we could see. A few deep crevasses scored the length of the berg. Out of every imperfection and gap in the white ice came an electric blue light that made the mist around it glow. The iceberg looked solid, immutable, a citadel of ice. But the light, dim as it was on this foggy day, had penetrated it, bounced around and emerged from this enormous object to create something ethereal.

It took the best part of ten minutes to sail past the iceberg, after which it disappeared at our stern, incongruously fast for something so big, into the fog.

Despite its size and solidity, it was clear that the berg was being slowly eaten away by the water in which it floated, no doubt for hundreds of metres more below the surface than any of us could see. On a human timescale it would probably still exist for a long time yet, but this iceberg, impressive and breathtaking as it was, was just a remnant, the dust and ashes of something much bigger, something alive that lay ahead.

Our enormous piece of ice had snapped off a glacier that had been built from thousands of years of snow falling onto the frozen continent. That glacier had flowed, over the course of thousands more years, to the edge of the Antarctic landmass and entered the sea. The iceberg, a tiny speck of that unimaginably huge glacier, had calved off into the water and floated away into the Southern Ocean. When we came upon it, this ice was already older than human civilisation and would probably encounter more ships in its remaining years. Scores of kilometres from the landmass of the continent, this was our first sight of a piece of Antarctica itself.

The day after the iceberg, the fog lifted to reveal the bluest skies so far of our week at sea, with just a few wispy cirrus clouds here and there. The Sun was high and strong.

The ship had dropped anchor among the floes and, since it was possible to see much further than the day before, I spotted several icebergs around our ship, the closest of them just a few hundred metres behind, others many kilometres away. Each of them giant, irregular boxes of ice, bright against the sky, as big as city blocks. Everything was still.

The day came, a week into our voyage, when we were allowed off the ship for the first time since stepping on board at Bluff. As much as it was an opportunity to see the ice close up, this morning's venture was about getting used to the frigid temperatures and windchill of being outside in the Antarctic for extended periods of time. It was a chance to practise the precise way you need to dress in these parts. Flinging on clothes, without thinking, for a trip outside in the polar regions is a dead-cert for frostbite, hypothermia or worse. Too few clothes and you won't be warm enough; too many clothes and you'll sweat. Not only does that mean you're losing water in an arid place, the sweat will freeze into your clothes and make you even colder.

I put on four layers up top – an insulating merino wool base layer, a merino wool jumper, a fleece designed to be extra warm but breathable and a down-filled coat – and three layers over my legs. I tucked my jumpers into trousers, sleeves into gloves and scarves into coats. Two pairs of socks went on next and I topped off the whole ensemble with a layer of Gore-Tex waterproofs top and bottom, an insulated hat and a lifejacket. I covered my face and neck in high-protection sun screen – it's cold in the polar regions but the Sun is also strong and direct, making these areas a deceptively easy place in which to get badly sunburned. Two layers of gloves went on last and then I made my way to the starboard side of the ship, where the crew were lowering one of the gangplanks into the water. Waiting to meet it by the side of the ship was one of the Zodiacs, a heavy-duty rubber dinghy whose buzzing engine was the only sound aside from the chatter of a few of the expedition members. It would take around a dozen passengers away from the ship and enable us to cruise among the ice floes, perhaps even allow us an up-close inspection of the nearest iceberg. Two other Zodiacs were gliding in circles in the water nearby, ready to pick up their loads when the first one had set off.

I stepped down the gangplank and, as I got to the bottom, one of the people already on the Zodiac reached out to grab my forearm. I held his in return and he pulled me on to the dinghy, which moved in the water as I made footfall. Around ten others were already seated in rows along the two sides of the boat, facing each other. When we were fully loaded, the driver, standing at the back of the boat with one hand on the throttle, kicked the engine into gear and we left the ship behind.

We were soon surrounded, up close, by clear, cold, dark blue water. No one spoke for the first few minutes as we felt the sunshine and wind on our faces and listened to the water rushing past. The scene looked tropical, though the air temperature was below zero Celsius. Any time I took my gloves off to take a picture, my exposed hand would quickly stiffen and need to be buried inside a warming coat pocket for a few minutes before it resumed any useful function.

Up close the ice took on many more forms than we could have discerned from the distance of the ship. The ice floes and icebergs we knew but between the biggest pieces were giant boulders of ice – bergy bits – and tiny clouds of greyish icy rubble suspended in the water, known as brash ice.

All of it was a different shade of white. The compacted snow on the surface of the floes dazzled in the sunshine as if it had been laundered in bleaching detergent. The ice underneath, also white, took on a bluish hue. Where the floes met the water, the ice glowed with a translucent aquamarine.

Against this assault of white, the water looked black. Mainly because it is so difficult to keep your eyes adjusted to the extreme contrasts in a single scene. Focus for a while on the water, and somehow blank out the surrounding ice, and you could discern rich, deep blues and purples in the waves. Look closely at this black and white world and all of the colours are there, glistening and glittering, coming and going across surfaces of ice and water as the light across them shifted.

The Zodiac moved between the floes and came up close to a trio of Adélie penguins resting on one of them. They were a mere few metres away, but largely uninterested in the boatload of people that had become fascinated with them. A crabeater seal lazed out in the sunshine on another floe and, every so often, a delicate snow petrel – pure white apart from its tiny black eyes – would swoop past. When we got some distance from the ship, the driver turned off the Zodiac's engines and we were silent. No ship noise, no conversation, no audible water movement, no wind. The only sounds were a call or the occasional scrape of snow as nearby penguins moved around.

A few minutes later the engines crackled back into life and we headed straight for one of the ice floes. Our driver revved the motor hard and told us to hang on to the Zodiac – he was going to try to land the boat on the ice, a piece of floe he had seen and thought would be strong enough for us all to walk around on.

The Zodiac hit the floe and slid to a halt. One of the crew at the front jumped off to see if the ice really was strong enough and then helped us, one by one, off the boat.

It was a relatively small floe, irregular in outline but about a third of the size of half a football pitch in total. It was at least a metre and a half thick, topped with a good helping of compacted snow. All in all, very stable. Two more Zodiacs saw what we were doing and manoeuvred themselves up alongside our ice floe. Several dozen more expedition members soon hopped onto our new land, ready for discovery.

We were all laying claim to a new territory here, a piece

of frozen ocean that no one had ever stepped on before. This 'land' had not existed a few summers ago and it probably wouldn't exist in a few summers' time. In its temporary life, there was almost no chance that any other human would come anywhere near it. It might sound ridiculous when reading about it now but, in the moment, I remember being wedded to the idea that this was our ice floe and that we had landed with a purpose – to explore and claim it on behalf of humanity. This ice had never been seen by human eyes or touched by their fingers and here we were, doing exactly those things, in the shadow of those heroic explorers of the past. (In intent if not anything else – our lives at sea so far had been a great deal easier and better resourced than anything the explorers of the Edwardian era were able to enjoy.)

We stomped around on the ice, lay on it, sat in quiet contemplation on it. Some of us got excited and threw snow-balls and took pictures. Others wanted to try to be alone with it, walking away from the group and turning their backs on the other expedition members in order to stare out towards Antarctica, unseen but there, 160km away over the horizon. All of us, in our own way, felt that sense of elation, awe and intrigue – again just a shadow of the real explorers of the past – that must come with setting eyes upon something on our Earth for the first time.

The floe became a special place almost instantly, familiar and hard to leave. Eventually we piled back onto the Zodiac and slowly picked our way back to the ship, circling several more ice floes, stopping by more groups of penguins and seals along the way. A few hours after we claimed our new land, the ship's engines kicked back in, the anchor was hauled up and we continued on towards the frozen continent.

We reached the Antarctic coast just over a week after leaving New Zealand, around three days after first sighting sea ice. Our ship cruised along the shoreline near the frozen continent, just a few tens of metres from the edge of the ice. The ice was part of a huge sheet stuck fast to Commonwealth Bay. Normally this area would have been open water but ever since a giant 78km-long iceberg called B09B had grounded itself in the vicinity several years earlier, the sea here had remained frozen. Every year, the ice got thicker and the solid reach of the continent grew slightly.

It was late in the day – around 5 p.m. – but the Sun was high. The air was still; the dark water was flat and calm. This might have been a tropical beach, save for the landscape covered in snow and ice and total lack of trees or any other plants. The only living things we could see were groups of penguins, hundreds of them, along the shore.

There being no convenient harbour around, the only way for the captain to stop the ship next to the ice sheet was to rev up the engines and ram his vessel into it. Which he did, several times and at several different locations. Each time, as the ship slowly recoiled from the ice, we could see the V-shaped wedge that it had left on each collision. Any nearby penguins scurried away out of reach. In return for all this, the ice took a payment too, lifting off large flakes of blue paint from the ship's hull.

Commonwealth Bay in East Antarctica is a historic part of the continent, near the place where the great British–Australian Antarctic explorer and scientist Douglas Mawson had landed in 1912, when leading the Australasian Antarctic Expedition (AAE). Mawson was ambitious and an innovator. He had declined an invitation to accompany Ernest Shackleton's return to the Antarctic in order to raise a considerable sum of money and fund his own journey. He was the first to send back daily wireless signals from the Antarctic and the first to take an aeroplane to the continent. His expedition was also heavy with scientific goals, with two-thirds of his crew scientists. Over the course of the expedition, his team collected two years' worth of data from the continent on everything from weather and wildlife to the Earth's magnetic field.

We had spent the past few days picking our way, sometimes frustratingly slowly, through dense patches of pack ice on our approach to the continent. Pack ice is the thicker floating ice cover that forms when chunks of sea ice get blown into each other. These can become enormous and, sometimes, too thick for the ship to easily move through. While our ship had been able to cut through the young, thin sea ice at lower latitudes, the pack ice from 65 degrees south had proved more tricky. For a few days, the captain had to patrol the edge of the ice, looking for leads through it that promised to take us to the other side. More than once we would have to go backwards, out of a lead that had initially looked good but had, instead, become a dead end of thick ice. Progress had become so slow that the expedition leaders had called a meeting to explain what was going on – it had been an exceptionally heavy year for ice and it might take us the best part of a week to get through it all and onto the continent. A sign of how quickly the environment can unexpectedly change around Antarctica came the next day when we had some luck: a south-westerly wind had shifted some of the pack ice around and the latest satellite maps showed clear paths to the continent.

Just over a day later, we were sailing through very different waters. The sea was dark and calm and there were no more ice floes dotting its surface. The only ice we could see were colossal icebergs – scores of them – all around us. Most of them were flat-topped mountains of ice with sides that glinted fiercely when angled just right against the Sun. Others had more interesting shapes – topped with sharp slopes or generally amorphous and organic-looking, free of most of the sharp edges you might expect from giant cubes of ice. Everyone on board the ship had different descriptions of the shapes. Some people just saw ice and marvelled at the forms into which this substance could contort itself given time and pressure. Others described collapsed buildings, the ruins of civilisation, cathedrals and minarets in and on the surface of the bergs, echoing the explorers of previous ages in their accounts. Contemporary concerns have always somehow been reflected in these inorganic cities of ice.

We spent a day sailing past an iceberg on one side of the ship and then another on the other. With a few of them, we got within mere tens of metres. Again, we saw translucent, electric blues in the cracks and curves that ran along their sides. But sailing past so many of them allowed us a glimpse into the sheer diversity of these behemoths. Inside some of them, coursing out of their imperfections, were greens and even violets.

Like the first one we had seen, these icebergs were all pieces of Antarctica on their way out to melt, eventually, in the Southern Ocean. But, as we sailed between then, we couldn't help thinking that they were a kind of honour guard, sent out by the frozen continent as a greeting, guiding us towards some grand, ancient civilisation. They were also a warning. Like their parent continent, these beautiful mountains of ice were also dangerous, inexorable, mostly unseen and unknowable. Down here the forces go beyond anything humans normally experience and the icebergs – giants that had been expelled from something greater than themselves – were a ready reminder of that.

When the anchor had been reeled out, the crew unloaded equipment and vehicles onto the ice sheet against which we had stopped the ship. Once an advance party had spent some time checking that the ice was safe to walk on, we trooped down the gangplank and onto a new continent.

The ice was covered in half a metre of soft snow and

walking around took a little getting used to. The ground, perfectly white and flat, stretched out to mountains on the horizon to the south. Somewhere beyond there lay the polar ice cap and the South Pole itself. To the north, icebergs were the only things visible on the sea.

For a while, I sat among a group of Adélie penguins at the water's edge, by far the closest I had ever come to these creatures without the glass of a zoo between us. I watched them waddle around, jump in and out of the water, slide around on their bellies. They didn't seem to mind having interlopers in their midst but they kept a distance. Even the ones that looked as though they were resting had a sixth sense that would allow them to wake and slide away in time if any non-penguin tried to get too close.

We got back on board the ship at midnight though, bustling with stories and photo-swapping. With the Sun still shining outside, it might as well have been first thing in the morning.

The Antarctic is covered by 10,000,000,000,000,000 (ten thousand trillion) tons of snow and ice over its entire land-mass, with tendrils and ice shelves that reach out from the main bulk of the ice sheet and into the Southern Ocean around it. As such, it is the largest single mass of ice on the planet and makes up a big chunk of the cryosphere, the frozen part of the Earth's water system. The ice is not an inert substance, a mere storage medium for water in places where the temperatures fall well below freezing – it is key to the entire climate, geology and life of Earth.

A quarter of the Earth's ocean area is covered in ice, which floats on the surface (an anomaly for sure, but a deeply important one in the history of the evolution of life on Earth). Sea ice alters ocean currents, it limits the exchange of gases between the sea and the atmosphere and its colour reflects the vast majority of the Sun's incoming radiation, regulating temperatures and keeping our planet cooler than it otherwise might be. In the air, ice crystals form clouds and act as sites that bring together chemicals so that they can react – the basis of atmospheric chemistry such as the one that destroys ozone.

The word 'cryosphere' comes from the Greek word for cold, *krios*. The largest parts of the cryosphere on land are the vast continent-sized ice sheets on the Antarctic and Greenland. Elsewhere in the world, land ice comes in the form of ice caps, glaciers, snow and permafrost. Around and in water, low temperatures can freeze the surfaces of river, lakes and oceans.

The Arctic is composed of sea ice, which shrinks and grows with the seasons, and glaciers and ice caps on the Canadian Arctic Archipelago, Greenland, West and East Arctic Islands, Iceland and Svalbard. Ice covers around 275,500 square kilometres of the Arctic, more than half of which is contained in the glaciers and ice caps on the 36,000 islands of the Canadian Arctic Archipelago. Another quarter is on the Greenland ice sheet, with much of the rest distributed around the volcanic ice caps of Iceland, and the mountains and fjords of Svalbard. In addition, the Arctic Ocean is covered in sea ice, which shrinks and grows with the seasons and often reaches ground on one or more of the coasts within which it is enclosed. This region's ice cover first showed hints of becoming perennial around 13 million years after the Earth was cooling down following a greenhouse period, according to scientists at Ohio State University who looked through evidence from dropstones to whale bones in an attempt to work out its history. A few million years later, the sea ice would disappear during warmer months but, for the past 2 million years or so, the ice pack has been a permanent fixture of the Arctic Ocean. For a large part of that time it would have been several hundreds of metres thick in many parts and, save for an exceptionally warm summer here and there when the ocean might have been ice-free, most of the formidable ice cover stayed in place.

At the other end of the world, the Antarctic, the ice-covered area is around 77,000 square kilometres, the majority of which sits in the Antarctic Peninsula and the Antarctic Ice Sheet. Surrounding the continent are the Sub-Antarctic islands including the South Shetland Islands, South Georgia, Heard Island and Kerguelen, with a total estimated ice cover of roughly 7,000 square kilometres.

All of this snow and ice comes in several forms, each made differently, each with its own particular life-cycle and effect on the planet.

The most familiar is the precipitation we know as snow, made from ice crystals in ice clouds that are below zero degrees Celsius. Once it has formed in a cloud, a snow crystal will absorb more water molecules from the air around it, growing until it is heavy enough to fall through the air and reach the ground. Sometimes the ice crystals fall through clouds of supercooled water droplets on their way to the ground, water that is colder than freezing point but which has remained liquid. These droplets, disturbed by the falling ice crystal, will freeze onto it and the resulting soft, crumbly mass is known as a graupel. If normal water freezes onto the ice crystal as it falls, we call the frozen mixture sleet.

Ice seems straightforward enough – water that has fallen below freezing point and solidified – but this substance has hidden dimensions, reflections of the complexity of its liquid counterpart.

One afternoon near Cape de la Motte on the East Antarctic coast, the captain lowered the gangplank and let us wander onto the ice next to our ship. In the hours before, the scientists on board had cored holes through the ice to ensure it was thick enough for us to walk on safely and, as we trudged onto the snow-covered ice, one of them handed me one of the slim tubes of ice he had pulled out of the ground moments earlier. 'Taste it,' he said, beaming. It hadn't occurred to me to do anything of the sort but given that this was a special piece of ice, I put a corner of the tube into my mouth and gingerly ran my tongue across it. It was cold, though that wasn't the biggest surprise: this ice was salty. That told the scientists, unmistakably, that we were standing on relatively new sea ice.

It seems counterintuitive, of course, that ice would be anything but crystals of pure water. But when sea ice forms, it traps concentrated droplets of brine in the pockets between the ice crystals. As the ice gets years older and thicker, the brine will eventually drain through the structure, leaving behind the pure, tasteless ice that we are all familiar with.

Sea ice is frozen ocean, forming, growing and melting while out at sea. It is a slow process because the salt in the water lowers its freezing point to around -1.8°C and, because oceans are so deep and can keep their temperatures relatively stable for long periods, it can take a long time to reach the correct temperature. In general, the top 100–150m of water needs to be below -1.8°C in order for the surface to freeze.

Despite these limitations, sea ice can form as far south in the northern hemisphere as Bohai Bay in China (38 degrees north) and as far north in the southern hemisphere as Macquarie Island (54 degrees south). It is highly seasonal, growing in the winter months and receding in the summer. Around 15 per cent of the world's oceans are covered by sea ice during part of the year.

This ice has an important effect on the amount of Sun's energy that reaches the oceans. Its bright surface reflects into space most (80–90 per cent) of the light that falls on it, meaning that ice-covered regions of the sea can remain relatively cool. This also means that, as global temperatures rise and sea ice dwindles, there are more dark spots of ocean to absorb sunlight and the cycle of warming becomes exacerbated, leading to warmer seas and more ice melt and so on. Even small changes in temperature can have big feedback effects on ice and, as such, the polar regions are the most sensitive regions on Earth to climatic changes.

The formation of sea ice around the poles is also a driver of some of the world's biggest ocean currents, the Global Conveyor Belt, as we have already seen (see chapter four). As ice forms on the surface of the water, it drives out the dissolved salt there, which leaves the water directly below the ice much saltier. That dense, salty water sinks to the bottom of the ocean and moves along the floor towards the equator. This current of water – known as Antarctic Bottom Water – pulls in warmer water from the tropics behind it.

Whereas sea ice is made only from salty ocean water, on continents, ice forms from fresh water and can appear on top of a lake or river during winter or on land in the form of snow caps or glaciers.

Glaciers are thick masses of land ice that are built up over hundreds or thousands of successive years of snowfall which, if it remains in place long enough, can be compressed into ice. As the years go by, the snowflakes deep in the glacier crystallise into sugar-sized grains, which get larger as more snow piles on at the top. The air pockets between the grains get smaller, the snow becomes more compact and turns into 'firn', an intermediate state between snow and ice that is around two-thirds the density of water. Over centuries, the firn turns into ice and the crystals push out virtually all the air, growing to several centimetres in length.

Glaciers cover around 10 per cent of the world's land, most of it in mountainous alpine areas or around the poles, and store 75 per cent of the world's fresh water. Many of them feed the rivers and lakes upon which billions of people depend for drinking and watering their crops. The ground-up rocks and soil left behind when a glacier has retreated is also valuable for agriculture.

Several of the great rivers of South-East Asia – the Yangtze, the Ganges, the Yellow, the Mekong and the Indus – are fed by the melt water from the more than 37,000 glaciers of the Tibetan plateau. Often called the 'third pole' of the world, this region contain the largest volume of ice outside the Arctic and Antarctic. In South America, the citizens of La Paz rely on the melt from a local ice cap glacier to provide water during drier spells in the year.

Glaciers can be long- or short-lived, relatively speaking, and can be as small as football pitches or hundreds of kilometres from one end to the other. The biggest glacier in the world, the Lambert-Fisher Glacier in Antarctica, is 400km long and up to 100km wide at points. More generally, the Antarctic is covered in many enormous glaciers, some of which are up to 4.7km thick in many areas.

What marks a glacier out from a bog-standard field of ice, however, is that a glacier inexorably moves downhill, pulled by gravity. The sheer mass of the ice inside them causes the edges to bulge and spread; meanwhile, the pressure of the glacier's weight melts a layer of water at its very bottom where it touches the ground, lubricating the whole thing's slide over the rocks and sediments there. Cracks in the ice can also allow water to trickle through to the bottom from the warmer top of the glacier. Either way, the ice at the top often tends to move more quickly than the friction-curtailed bottom, deforming the shape even further.

Depending on environmental conditions – how much snow they accumulate versus the amount of evaporation or melting that occurs – glaciers can advance or retreat, which means that the tip of the glacier moves forwards or backwards. This is a very slow process, often taking years or decades, but, once in a while, glaciers have been recorded as surging, where the movement can be spotted over a period of months. In 2014, scientists at the University of Washington in Seattle reported that the movement of the Jakobshavn Isbræ glacier in West Greenland, which has been retreating since the mid-1990s, had sped up by 30–50 per cent in the previous two years. In the summer of 2012, they reported that the glacier was moving at the record speed of more than 17km per year – four times greater than during the 1990s and faster than any sea-bound glacier ever measured in Greenland or Antarctica.

Over hundreds and thousands of years, these vast rivers of ice consume rocks, villages, people. They pick up everything in their path and, over time, move them to new places – in studying tracks of old glaciers to work out how they moved, geologists often look for 'erratics', rocks that seem out of place where they are. These rocks have literally been taken from their source and dumped somewhere else, ending up out of place, by the moving ice that has subsequently melted.

Over the ages, glaciers have carved out mountains and valleys in their wake. The remains of glacier tracks are visible all over the world – trough-shaped valleys with steep cliffs where entire mountainsides have often been carved open. The Yosemite National Park's deep valleys and the striking fjords of Norway were carved by glaciers. The Matterhorn in Switzerland is a result of several glaciers coming together to erode the original mountainside from all sides, leaving behind the distinctive pointed shape of the mountain.

When a glacier reaches a sea or a lake, chunks of it will float out onto the water until, eventually, some of the ice snaps off and floats away into the sea. These are icebergs. These gigantic pieces of floating ice can be anything from the size of a small car to a small state (Antarctic ice shelves can often produce icebergs that are over 80km long). They come in a myriad of shapes and a symphony of blues and whites.

Icebergs have important influences on ocean life. The

water around them tends to teem with plankton and fish because the bergs leak out nutrients they have carried off from the land. Since the majority of their volume and mass is underwater, these blocks of ice move with the currents in the ocean and go where the water sends them – out towards the tropics, into each other, grounding onto land. As they travel, sometimes for decades, they change shape as the air and temperature attacks their surface. Warm air creates pools of melted water at the top, which can trickle down through the berg to create cracks and widen into crevasses. As the iceberg moves further into warmer latitudes, the warmer waters erode the underside into wild shapes, melting and disintegrating it from the bottom. In its lifetime, the iceberg will bear the scars of these insults in its changing shape, its shedding and breakdown into pieces and, eventually, in its complete disappearance.

This process of disintegration is natural, of course, and scientists have found that watching it can be useful in working out how the ice cover at the poles will behave as the world gets warmer. The way icebergs break up is reminiscent, they have found, with the way Antarctic ice shelves (permanent floating sheets of ice that are connected to the continental land mass) break apart.

There is a classic science demonstration used by anyone on the border of being a scientist and a wannabe party entertainer. They will produce a bottle of cold water, normal-looking in every possible respect. They might hand it over to a member of the audience to inspect, asking them to gently move the water around inside to show that there's nothing abnormal going on. Taking the bottle back, the scientist/entertainer will then hit the bottle hard against a table or a hand. The water inside the bottle freezes solid, instantaneously. The audience usually gasps and cheers. Surely there's trickery involved, they'll be wondering. It can't be real. Can it?

In fact, it's very real and very easy to do. And, with apologies to any magicians out there, I'll explain how. The scientific ideas behind the instantaneous freezing in this party trick are a brilliant insight into one of the most familiar, yet badly understood, processes that occurs every second on Earth – how water freezes into ice.

As the bottle is cooled down, the water molecules within become more sluggish. Below 0°C, the most stable configuration of molecules, thermodynamically speaking, is in an ice crystal. To do this, the molecules need to move and bond together into the appropriate pattern for ice. This is not as easy as it sounds, though. If water is cooled quickly enough, the molecules might not have the time to move themselves into the correct arrangement for an ice crystal before they become too sluggish to move around very much at all.

Any ice crystal (in fact, any crystal) needs a point of beginning, a place of nucleation where the first molecules manage to arrange themselves into the correct structure and then other molecules pile on in the same arrangement to grow the crystal further. The moment when and how that initial nucleation happens, in order to start the solidification process, is something of a mystery.

'If you bang the bottle, you disturb the system and this disturbance then sets off some nucleation event,' says theoretical chemist Angelos Michaelides, who has spent most of his career working on how water freezes. 'But people don't know exactly what happens.'

The best way to prepare a bottle of water for the sudden-freezing demo is to place it into a bath of ice and salt, which will leave the water inside the bottle at around -10°C. Liquid water at this temperature does not exist normally and, since it hasn't undergone the expected phase change into ice, scientists say it is in a 'metastable' state. That means the bottled water is ripe to undergo a sudden phase change if nudged in the right direction.

Perhaps the bang on the bottle disturbs tiny impurities, dust particles for example, within the liquid and the cooling water molecules attach themselves in the right way to those particles. The disturbance might take the form of air that is normally dissolved in the liquid but which comes out of solution, as tiny air bubbles, when the bottle is banged. It might even be that a wrinkle on the inside of the bottle becomes a place where the water finds itself in the correct arrangement to begin nucleation.

Another version of the trick involves pouring the ice-cold water from the bottle onto a cube of ice. The spectacular result is that the water freezes as it touches the ice and the crystal grows before the awe-inspired spectators' eyes. The thinking here is that ice can nucleate if water at the right temperature finds a good template for ice. And what's a better template for ice than ice itself?

That we know so little about how water freezes might come as a surprise, so it's worth making a few definitions clear. Melting and freezing of water are so common as to be inconsequential in our lives, mundane, even. We think of them as simple commutative processes – freezing happens when you cool water down, melting happens the other way around.

But our understanding of these two processes is asymmetric. Ice will always melt at 0°C under ambient conditions, indeed the zero of the Celsius temperature scale is defined by the melting point of water. But 0°C is not the freezing point of water. Liquid water can freeze, under differing circumstances, at any point up to -38°C. The energy differences between the two phases – liquid and ice – at or around 0°C are so small that water can seem to hover between them when it is in the act of crystallising, seemingly at odds with the laws of physics. Whether or not ice forms in a given situation is one of the more important phase changes in chemistry. It can be the difference between life and death, and we have barely started to get a handle on what happens. There is more water melting and freezing than any other chemical process going on in the world and we still don't know the basic mechanism by which water freezes. 'I think that's a travesty,' says Michaelides.

Here's what we do know. Think of a water molecule in terms of a ball-and-stick drawing where two hydrogens are attached to a central oxygen, with the angle of around 104.53 degrees. These float freely around each other in the liquid phase, forming and breaking passing alliances with one or many other molecules due to hydrogen bonding, the relatively weak interaction where the small negative charge on one side of the oxygen atom attracts the small positive charge on a nearby hydrogen from another molecule. (As we have already seen, hydrogen bonding is at the root of many of liquid water's anomalies.) These weak bonds mean that each water molecule is ideally surrounded by four neighbours and this is the basis for the tetrahedral structure we see in ice crystals.

On the liquid side, we have hydrogen-bonded molecules vaguely forming the outline of a crystal and, on the ice side, the structure is rigidly set. What happens to go from one to the other?

'We're talking about a process called nucleation of one phase in another,' says Michaelides. 'That's one of the least-understood phenomena, in general, at the atomic and molecular scale.'

We know very little about nucleation events because, usually, they involve very small collections of atoms. To probe them requires a microscope that can provide high spatial resolution for processes that happen astonishingly quickly, on the picosecond timescale (10^{-12} seconds). This is a staggeringly small amount of time, equivalent to one second as a second is to 31,700 years. In a glass of water, we would want to establish precisely where the first stirrings of order, a loose agglomeration of water molecules called a crystallite, took place.

'Once we see it, however, it's already happened,' says Michaelides. In so-called 'classical' nucleation theory the basic idea is, if we take our substance and we cool it below its freezing point, then the substance is trapped in a meta-stable state; it wants to be frozen but it just can't yet get over the hump and start the crystallisation process. The molecules will often arrange themselves into a little crystallite but, since the crystallite is embedded in a liquid environment and it will be getting jostled by the other water molecules, it will be generally very short-lived. Only when the crystallite gets to a large enough size will it become stable and the crystal will start to grow.

And that is a random process.

What scientists such as Michaelides are looking for is something they call the 'critical nucleus', a little crystallite that is large enough to be stable when embedded in the water.

In supercooling liquid water, tiny little pre-critical nuclei will be forming and dissolving all the time. The size of these nuclei then depend on how supercooled the water is – when it is only a little bit supercooled, you need to form a very big crystallite before it's large enough to survive. But when the water is deeply supercooled, you only need a very small crystallite because it becomes more energetically favourable for the water molecules to form the crystallite in the first place and also because fewer other molecules are jostling it around to break it up. The energy barrier to forming a crystal reduces as the temperature goes down. There comes a point where you can't supercool water any further and that's when this energy cost to make the pre-critical nucleus disappears. That's when we get a spontaneous formation of ice and a homogenous freezing point of -38°C.

In computer simulations Michaelides has found that, if you supercool water to -20°C, the size of the critical nucleus might just be a few hundred water molecules. Go to higher temperatures and the cluster needs to get bigger as the driving force to form ice is reduced. Close to the melting point, -1°C or -2°C, the critical nucleus might be tens of thousands or hundreds of thousands of water molecules. This will be a major challenge to simulate.

Does any of this matter outside the vaulted academy of theoretical chemistry? Writing in *Nature* on the most pressing problems in our understanding of the behaviour of water, chemist Thorsten Bartels-Rausch said that how and when water freezes was critical knowledge for the under-standing of the Earth's climate and water cycle. Without it, he said, 'we cannot predict with certainty when and where ice clouds will form in the atmosphere; areas of the sky remain humid when we would expect them to freeze. Do water droplets freeze from the surface first or crystallise from within? Which form of ice will they make?'

One of the greatest uncertainties in the computer models for climate and weather is how ice forms in the upper atmosphere and how much of it there is up there at any given time. The atmosphere is filled with all sorts of contaminants – little clay particles swept up from deserts, soot, chemicals that come out of factories and cities; there's even bacteria up there. It is still unclear which of these nucleate the most ice.

Climate scientists need to build better climate models that can help them predict, say, how much ice will form in a part of the atmosphere given specific concentrations of chemicals that are present. This is crucial to know because the amount of ice in the clouds controls the amount of rainfall and the amount of sunlight that gets reflected back into space, all of which have big effects on the Earth's energy balance.

'The uncertainties of the role of ice in these climate models is much greater than any uncertainties in what we know about CO_2. The role of water and ice in these climate models is one of the biggest uncertainties,' says Michaelides. Scientists don't even know whether it leads to cooling or heating overall – its effect sits close to zero and the error bars straddle both sides.

Airlines want to know how ice forms on the bodies of their craft and whether there might be ways to prevent it – they already coat planes in anti-freeze chemicals to prevent ice crystals growing. Back on the ground, understanding how ice forms is useful if you have an oil or gas pipeline in a cold place, in case the ice blocks it and reduces production.

Airlines and oil companies both use petrochemical-based anti-freeze agents, which can disrupt the energy balance of the system enough to destabilise the formation of ice. But oil and chemical companies need a lot of this stuff – an offshore oil rig might have to store gallons of these flammable materials – and so there is an incentive to find chemicals that work more efficiently.

All of the ice known on Earth is made from rings of six water molecules, arranged in chicken-wire sheets and stacked on top of one another. This is the ice on Antarctica, the ice floating on the oceans, the ice that turns milk into ice cream and the ice that is the solid water you make in your freezer and later use to cool down your drinks. Chemists call it ice-Ih (the 'h' stands for hexagonal) and, for a long time, that's all they thought there was. Until now, you probably did as well.

In ice-Ih, which exists at the ambient conditions we experience on Earth, the water molecules come together in a relatively limited set of rules: each molecule has four others bonded to it in a tetrahedral arrangement (they are 'four-co-ordinated').

This arrangement leaves large open channels in the crystal, making it less dense than liquid water. This is the reason that ice floats. The multiple forms of frozen water we encounter on Earth – snow, ice, slush – are metaphorically reflected at the molecular level; they are phases in which the water molecules are bonded and linked in different ways. The simplest way to see the relationships between the various phases of ice is to look at a plot of temperature against pressure, known as a phase diagram (*below*). Different regions of the diagram correspond to the different conditions required for each crystal phase of ice. There are sixteen distinct phases of ice known to scientists – each a type of crystal with a different arrangement of water molecules – and that number

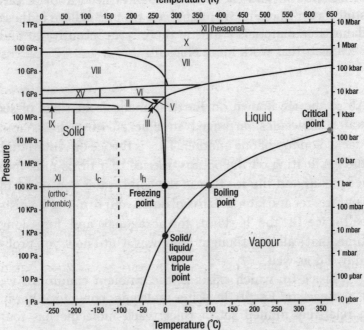

is certain to grow in the future. Though wildly different in many ways, they all still maintain the basic fourfold co-ordination rule seen in ice-Ih. To explore the stranger phases of ice, we need to add pressure.

'As we decrease the volume available by increasing the pressure, the water molecules still satisfy the fourfold co-ordination, but we have to allow a certain amount of bending of the bonds that link neighbouring molecules,' writes physicist John Finney. 'Different mixtures of ring structures result, with an increasing degree of bond bending as the volume available to the phase is reduced. We have to bend our basic jigsaw pieces a little to squeeze them in. As the available volume decreases further, a little lateral thought is needed: we can't continue to bend the bonds indefinitely – the jigsaw pieces might break – so we make use of the fact that we have open structures, and hence available volume through which to thread some of our hydrogen bonds. At the really high densities, we take this process to the extreme and end up with two interpenetrating lattices that eat up all the available volume and in essence give us packed structures.'

Squeeze ice-Ih to 5,000 times atmospheric pressure and you get ice-II. Though there are still open channels in the crystal, the hexagonal rings get squashed into smaller ones. Compress ordinary ice-Ih at -20°C and you enter the region of ice-III. Keep compressing and the molecules will arrange themselves into ice-V and then ice-VI and, you guessed it, ice-VII. (For those counting, we will come back to ice-IV shortly).

In Kurt Vonnegut's book *Cat's Cradle*, the world is threatened by a substance called ice-9, which is not only more stable than other ices, it is more stable than liquid water. A chunk of this substance could turn the water around it into more ice-9 and, in the story, it threatens to freeze the oceans of the Earth solid.

Real ice-IX does exist and it is not only 20 per cent denser than ice-Ih, it is more stable than ice-Ih and liquid water. But don't get worried – it can only survive at high pressure and extremely low temperatures. Bring it back to ambient pressure at -196°C and it will remain in the ice-IX form. If you dropped that into liquid water at room temperature, though, it would melt without any cataclysmic consequences. (At least that's what the theory says, though I'm not aware that anyone has actually done this for real.)

At higher pressures, the ice crystal deforms from its arrangement in ice-Ih. The open channels are not stable with increasing pressure and they begin to close up as the water molecules look for more efficient ways to pack together. Two things tend to happen as ice comes under pressure – the hydrogen bonds between the molecules begin to bend and water molecules that are not hydrogen-bonded move closer together. In ice-Ih, each water molecule has four hydrogen-bonded neighbours close by. But in the highest-density phase, ice-VIII, the crystal has been drastically squashed so that, in addition to the standard four hydrogen-bonded neighbours, each water molecule has also been crowded in by an extra four other water molecules that have been forced into its vicinity. In the highest-density phases, the networks of hydrogen-bonded water in the crystal end up overlapping with each other in order to pack together more tightly. In that regard, Ice-IV is one of the more interesting phases of the crystal made by scientists, since its structure is on the cusp between the open, spacious arrangement of Ice-Ih and the denser phases that appear at high pressure.

'As you go to higher pressures, the water molecules need to pack more efficiently,' says water chemist and ice specialist Christoph Salzmann. 'There are incremental ways to make them pack more efficiently. In ice-III, they're already packed a little bit more efficiently but not as effectively as in ice-VIII, for example. Experimentally, ice-X is the densest phase of ice found, though computationally they are predicting several denser phases of ice. All high-pressure phases of ice are denser than water.'

If you look along a hydrogen bond, one O-H distance is shorter and the other one is larger. If you keep squeezing ice-VIII, the hydrogen bonds will become symmetric, the two distances become the same and it becomes ice-X, which is sometimes called 'polymeric ice'.

The first high-pressure phases of ice were discovered at the start of the twentieth century. The naming convention, using Roman numerals, was established early on by Gustav Tammann and Percy W. Bridgman. More than 100 years later, Salzmann's team at University College London are still finding new ones, though it is getting increasingly harder.

Since 2006, Salzmann and his colleagues have made ice-XIII, the hydrogen-ordered counterpart to ice-V; ice-XIV, the counterpart to ice-XI; and ice-XV, which is the counterpart to ice-VI. (The numbers can get a bit confusing since they are chronological to a discovery rather than relating to any particular part of the phase diagram – when someone finds a new phase of ice they just give it the next number.)

So far, scientists have almost completed their discovery of the pairs of crystal arrangements. Ice-IV doesn't yet have an ordered counterpart but there's no doubt that it must exist. Conversely, the disordered counterpart to ice-II hasn't yet been made either. Though predicted by computer models, the laboratory equipment needed to create and measure these new high-pressure phases of ice does not exist yet. The new phases of ice are predicted to form at around 700,000 atmospheres (700 gigapascals, or GPa) and the record for the highest pressure achieved in the lab is around 200 GPa. Salzmann says it is a technology issue – whoever leads in creating the highest pressures in the lab will make a new phase of ice. In fact, whenever lab equipment manufacturers develop a new pressure kit, they will usually use it first to study ice. 'There is no doubt,' he says. 'We know there are more efficient ways of packing the water molecule so there must be more phases of ice.'

That is not to say that packing molecules closer together is the only way to make new phases of ice. The newest phase was discovered at the end of 2014, by Thomas Hansen at Institut Laue-Langevin in France and is in the form of something called a clathrate. A clathrate is an arrangement of molecules that looks something like a cage and usually it encapsulates some other guest molecule. Clathrate hydrates are cages made from water molecules and vast deposits of them are known to exist at the bottom of the ocean sediments and in oil pipelines, each water cage containing hydrocarbons such as methane. Hansen and his colleagues at the University of Göttingen wondered if the clathrate cages could be the basis of another type of ice – in other words, could the cages exist without anything inside them? To test this, he created clathrate hydrates containing neon atoms, which are inert, and then placed the whole ensemble inside a vacuum pump at -133°C. Over five days he pumped out the neon and found that the water cages remained intact, even though they ended up empty. Ice-XVI, as Hansen called it, has the lowest density of any known form of ice, at just four-fifths that of the more common ice-Ih. It is the first form of clathrate ice made in the lab and Hansen's work opens up the intriguing possibility that there could be many more forms of ice like this, different arrangements of water cages that can exist without any guest molecules inside.

There is one more form of ice that is worth mentioning, another form of the familiar ice-Ih and which, by rights, should exist on Earth naturally. Recall that ice-Ih is made from stacked sheets of water molecules, arranged in hexagonal rings. Alternate layers are arranged so that they are mirror images of each other. There's another way you can stack these sheets, where the hexagons are directly on top of one another – this is cubic ice, known as ice-Ic. The hexagonal arrangement is marginally more favourable from an energetic point of view but the difference is so small that it is easily within the resolution of any theories or experiments you can do.

It is theoretically possible to make ice-Ic in the lab. Take a high-pressure phase, say ice-XV, and heat it at ambient pressure and it should turn into ice-Ih. Before it turns into ice-Ih, though, it goes through a stage where cubic ice appears. Salzmann says this transformation can be seen happening in real time: if you take ice-XV, which is much denser than ordinary ice, and warm it up, you can see the phase transition as it happens, as the ice pops up and the volume expands.

Even so, ice-Ic seems to be very rare indeed.

'It turns out that perfect cubic ice has never been made yet; nobody has made an ice sample where you just get this cubic stacking,' says Salzmann. 'You [normally] get something we call "stacking-disordered ice". Whenever you make a sample, yes, there is a certain fraction of cubic stacking, but there are always hexagonal sequences in the material as well. The important thing is that this is stacking-disordered ice – that name describes a very broad range of possible structures. For example, if you look at hexagonal ice and put in a few cubic sequences, that would already be stacking-disordered ice. This material spans all the way from pure hexagonal ice to pure cubic ice and it really depends on how you make it [as to] what fraction of hexagonal and cubic ice you have. We're trying very hard to make pure cubic ice. We've got 80 per cent of the way.'

Making perfect cubic ice sounds like a curiosity-driven exercise and, for Salzmann and his colleagues, the challenge and

difficulty of doing it is part of the thrill. But, he adds, it is
practically important that we understand more about the
properties of this stacking-disordered ice and how the
stacking disorder affects the properties of ice.

Stacking-disordered ice is assumed to exist in the upper
atmosphere and in space and would have some intriguing
properties. Salzmann recently predicted the structure of an
ice crystal made from this type of stacking-disordered ice and
found that it wouldn't have the six-fold rotational symmetry
of something like regular ice-Ih but, rather, a three-fold
symmetry. Every now and then people find snowflakes with
three-fold rotational symmetry and Salzmann thinks this is
evidence for the existence of stacking-disordered ice in the
upper atmosphere.

If this exists in the upper atmosphere as Salzmann
suggests, it could have important effects on how climate
scientists incorporate this ice into their computer models.
'First of all there is an energetic reason – when it transforms
into hexagonal ice there will be an energy release, which
will have an impact on the climate,' says Salzmann. 'The
light-scattering properties of stacking-disordered ice will be
very different compared to hexagonal ice. Which will also
have a huge impact on the climate as well if the incoming
sunlight is scattered differently depending on the type of ice
you have up there.'

In fact this light scattering is another potential way to
remotely spot stacking-disordered or cubic ice in the upper
atmosphere. 'Sometimes people see these haloes around the
Sun – when it shines through an ice cloud – and cubic or
stacking-disordered ice would be a possible explanation for
it,' says Salzmann.

The energy difference between stacking-disordered ice
and hexagonal ice is not much, but multiply it by the vast
amount of ice there is in the atmosphere, and it becomes a
significant value in terms of the energy in that part of the

Earth's climate system. Apart from the possibility of ice-Ic somewhere in the clouds, the only other phase of ice that might exist outside the lab is ice-VII. A group of geologists proposed that it might be present in a cold sub-conducting slab in the bowels of the Earth, where tectonic plates move underneath other plates. It'll be hard to prove, though.

In the real world, the only phase of ice we know definitely exists on Earth is ice-Ih. Apart from ice that has been created in the rarified world of the lab and some ideas based in clouds or in the earth's mantle, the only other places where scientists think that the higher-pressure phases might exist are on other planets, where the surfaces or cores might contain water held in very different circumstances to that on Earth.

We will come back to the possibilities of water on other planets later in the book. But now I want to finish with a thought experiment into what remains to be discovered about the way water can freeze – in other words, what other ices might exist, if we could somehow simulate them?

There will be more phases of ice at ever-higher pressures, as the water molecules continue to get squashed into more efficient patterns and become more ordered in their orientations. But that seems normal by now.

Imagine a perfect packing of the molecules, something that cannot be compressed much further without the atoms themselves being ripped apart – in other words, something we could still call water. 'That is basically a close-packed structure,' says Salzmann. 'If you take oranges and you start packing them, you make a hexagonal arrangement in the layer and you put the oranges where the gaps are, this has been shown mathematically to be the most efficient packing of spheres. There is no denser way of packing spheres. For ice you could just think that the oxygen atoms are represented by the oranges and the hydrogens are so small that they will just go somewhere in the gaps – this will be the densest phase.'

This is still hypothetical but, if this phase exists anywhere, it is likely to be somewhere in the centre of a star or a gas giant planet such as Jupiter, where the pressure is epic, well in excess of 1,000 GPa. 'Potentially it could be metallic, with the H+ acting as the charge carriers. If you have a metal or not, that's an important question to clarify. If you have a metal, most likely you'll have a magnetic field associated with it, ring currents and everything.'

Ice as a metal? The water molecule can certainly do some surprising things. But for now let's head back to ice as we understand it in our world, back to Antarctica.

EIGHT

Cape Denison

In the first week of January 1912, Douglas Mawson sailed into what he had named Commonwealth Bay, part of a territory in East Antarctica that had already been named Adélie Land. His first task was to look for a place to set up his main base, so he cruised the coastline for several days. Writing in his account of the expedition, *The Home of the Blizzard*, he recalled a rigid, inhospitable-looking environment:

> The land was so overwhelmed with ice that, even at sea level, the rock was all but entirely hidden. Here was an ice age in all earnestness; a picture of Northern Europe during the Great Ice Age some fifty thousand years ago. It was evident that the glaciation of Adélie Land was much more severe than that in higher Antarctic latitudes, as exampled on the borders of the Ross Sea; the arena of Scott's, Shackleton's and other expeditions. The temperature could not be colder, so we were led to surmise that the snowfall must be excessive. The full truth was to be ascertained by bitter experience, after spending a year on the spot.

They could not have known it at the time but they had arrived at the windiest place on Earth. They would get average wind speeds of 80kph here, with gusts up to 250kph. These are the fearsome Antarctic katabatic winds, frigid air pulled down by gravity from the polar plateau behind Commonwealth Bay and funnelled out to sea. The title of Mawson's book, *The Home of the Blizzard*, was an apt description for the location.

In that first week of searching for a base location during January 1912, though, Mawson and his expedition team did not know any of this. In fact, by 8 January, their search had led them to a bay covered in thousands of Weddell seals and Adélie penguins. Sailing towards a small islet in the rocky bay, they found themselves 'inside a beautiful, miniature harbour completely land-locked. The Sun shone gloriously in a blue sky as we stepped ashore on a charming ice-quay – the first to set foot on the Antarctic continent between Cape Adare and Gaussberg, a distance of 1,800 miles.'

They named the rocky area nearby Cape Denison and decided to build their main hut opposite Boat Harbour, into which they had sailed. The original plan had been to build a main base and two satellites further along the coast in order to record a wide range of geographical and meteorological observations. The density of the pack ice they had encountered on their arrival at the continent, however, had scuppered that plan. Instead, Mawson integrated the smallest of the three bases into his main team and sent a secondary team of eight men, under the leadership of Frank Wild, to set up a second base further along the coast.

Over the next few weeks, suffering bitter cold and driving katabatic winds that often meant they could barely stand up straight, Mawson's men built four wooden huts at Cape Denison, each with a specific purpose.

The largest hut contained living quarters and a workshop.

Next to that was the Transit Hut, used to make astronomical observations. Further down the hill towards Boat Harbour were Magnetograph House and the Absolute Magnetic Hut, where Mawson's team studied the variations in the Earth's magnetic field.

The main hut was a square-framed box with a pyramid roof, which ended with five-foot-high walls on all sides. Around three sides of the hut was an enclosed verandah, used for storage and also to provide a little extra insulation for the occupants inside the main space. Mawson's expedition team would stay here for two years.

A century after Mawson had arrived, Cape Denison was a different place. In 1912, Mawson sailed his ship right into Boat Harbour, barely half a kilometre from the eventual location of his huts and, until a few years ago, any modern visitors to Mawson's huts could have done the same.

By 2010, though, the iceberg B09B had got in the way – 78km long and 39km wide, it had originated somewhere to the east near the Ross Sea and swept along the Antarctic coastline, knocking off a 100km length of the Mertz glacier tongue in the process. The remnants of the ice tongue broke up and then dissipated around several places along the coast while B09B, the cause of the trouble, grounded itself at the mouth of Commonwealth Bay.

The upshot of this game of Antarctic billiards was that the entire sea around Cape Denison had begun to freeze. The katabatic winds that tumble down the polar plateau behind Cape Denison freeze the sea surface around the bay, but the moving air usually also blows this fresh ice out towards the open ocean, leaving the Boat Harbour area clear. With B09B in the way, though, the new ice had become stuck in place and, in essence, had extended the Cape Denison coastline far out into the ocean. For several years, scientists had waited in the hope that B09B would move on and that the ice might dissipate. Instead, it had just thickened.

Satellite maps of this area of fast ice – so-called because it was stuck fast to the land and was relatively stable – showed that it was three metres thick across most of its size and four or five metres thick in some places where it had stacked up or been compressed. This fast ice was growing in area every year and getting thicker. By the time we got to Commonwealth Bay in December 2013, the closest we could get our ship to Boat Harbour was at the edge of the ice sheet, around 70km away.

Given that we could not sail the *Akademik Shokalskiy* to the location of the huts, our expedition leaders had instead come up with a plan to drive across the fast ice in open-topped amphibious buggies. Made by a company called Argo (confusingly the same name as the deep-sea scientific instruments we encountered earlier, but definitely no relation) as a cross between a jeep and a boat, the eight-wheeled buggies were well equipped for such a journey and widely used in polar climes. The idea was that, if the ice underneath cracked or melted, the Argos would float, rather than sink into the frigid 500m-deep water below.

The first few days after we arrived at the edge of the fast ice, the scientists and expedition leaders on the ship spent a while scouting the nearby ice, to try to work out how thick it really was and whether or not there would be a safe path across it for the vehicles. Satellite maps of the area suggested that the fast ice was certainly thick enough for vehicles over large areas, but these maps were not finely resolved enough to show whether or not the surface was flat enough to drive over.

Their initial scoutings had been less than positive. Though much of the terrain was flat, there were also jumbled blocks of ice everywhere, ridges rose where the edges of former ice floes had been compressed together and there were cracks where tidal movements had broken the ice sheet apart. On top of all that, a recent warm spell had melted parts of the fresh snow cover on the fast ice, leaving behind pools of water covered by thin, fragile ice when the pools had temporarily refrozen overnight. At best, this meant any journey across the ice would be painstakingly slow, as we would need to keep a keen eye out for potential traps along the route. At worst, one of the Argos could get stuck in the slushy snow or in a pool of water, leaving its occupants stranded, possibly without any chance of help, for a day or more.

Operating in Antarctica is full of these sorts of unknowns. The environment – sunny, clement and peaceful in the morning – can change drastically by the early afternoon to blizzards and zero visibility. Weather forecasts can help but there are still no guarantees. You might have clear maps of a fine surface, taken from space with the best imaging instruments humans can afford to send there, yet the ground truth might be somewhat different.

Ice sheets are also prone to breaking up quickly, as the environmental conditions change. On the evening that we had arrived at the edge of the Antarctic coastline, for example, we had watched from the observation deck of the ship as, in less than half an hour, many kilometres of sea ice began cracking off from the edge of the fast ice and drifting into the ocean. Less than an hour earlier, the entire ice sheet had looked solid, stable for as far as the horizon at least.

For most short trips, sea ice that is at least half a metre thick is safe. The further you need to go, so the risk of finding and driving over a weak patch in the ice increases. The satellites had suggested that the fast ice near B09B was around three metres thick but, during the scouting exercise, the expedition leaders had drilled through it and found it was more like two metres thick in most places. Still safe for the Argo vehicles but a salutary warning that everything needed to be checked at the location rather than relying on distant measurements.

In addition to the several surveys of the ice, the expedition leaders had put in place multiple lines of communication between the Argos and the ship (and beyond). They carried three different models of satellite phones with them, and took VHF radios with built-in GPS and a satellite data system that they could plug into their laptops to send a message, if everything else failed.

The plan was that two Argos would make the trip while a third would remain on board the ship, in case it was needed for search and rescue. If the expedition team became lost and the ship had not heard from them at the pre-arranged times, the ship could issue a call for help from the nearest Antarctic station, the French base Dumont d'Urville, which was a few hundred kilometres around the coastline from Cape Denison. In an emergency, that base could scramble helicopters to assist in a search operation.

With the preparations done, multiple safety and communications protocols memorised, scientific equipment, emergency food, clothes and temporary shelter packed, the only thing left was to try out the route.

The first team of six went without much fanfare, perhaps an attempt by the expedition leaders not to draw too many false hopes from everyone else on the ship. They spent almost six hours making the 70km journey from the ice edge near our ship to Mawson's huts. It had been a hard slog, starting at 6 a.m. to try to get across the ice before the Sun was too high in the sky and causing more problems by melting the snow in the path. Still, the journey involved regular stops to pull their vehicles out of the sludgy surface of the fast ice. They faced big blocks of broken sea ice and deep snow. They had to backtrack several times, keeping their progress slow and careful, in the knowledge that, at any stage, all their routes forward might well be blocked; their hours of searching for a route to the huts might be for nothing and they would have to come back to the ship without completing their journey.

The original plan had been to pioneer a route to the huts and then take the remainder of the expedition members, in several trips, back and forth from the ship. But that first trip across the ice put paid to such plans. The expedition leaders decided it was just too risky, too arduous a trip to take too many more people without the chances of being stranded or encountering weather problems becoming too great.

By the time the first team reached the huts and started to carry out the scientific measurements on the ice and rocks there, they had also decided, during satellite-phone consultations with the expedition leaders back on the ship, to restrict any further trips to the huts to just one. After twelve hours at Cape Denison, just as the Sun was lowest in the sky, the first team made their way back, following the GPS locations they had marked out, like a trail of breadcrumbs, on their way in.

The first I knew about any of this was when the expedition leader who had remained on board the ship came into my cabin at 10 p.m. on a Thursday night and asked me if I wanted to go to Mawson's huts. If I did, I needed to be packed by 5 a.m. the next morning, when the first team were due back and the crews would be changed over.

A few hours later I was ready. There were five of us in total on the second trip, including two marine biologists who wanted to study how the fast ice sheet had affected the organisms living in once free-flowing, shallow waters around Cape Denison, and an ornithologist who wanted to record bird populations in the vicinity of Cape Denison.

Once the first team had returned, reddened from the Sun and tired, it took some time to get the two Argos refuelled and reloaded. In the back went spare fuel, our bags (packed with a complete change of clothing in case anyone fell into water and had to be warmed up rapidly), survival kits including tents, flares and blankets, and a small amount of scientific equipment. Reducing the weight on the vehicles was important – the less stress we placed on the ice, the less likely we would be to break it on our five-hour journey across it. The scientific equipment had had to be stripped down the previous night to take only what was absolutely necessary, the bare bones for the required measurements.

By the time I made my way down the gangplank to the awaiting Zodiac that would take us the short distance to the ice edge, it was 6 a.m. and the Sun was already bright in the sky. The air was fresh but not at all cold. This was good for our general comfort but not necessarily ideal for our proposed journey – the more Sun there was, the more likely we were to encounter water and slush on the ice during our journey. The temperature had been well below zero the night before and that gave us the best chance that the surface of the ice would be as solid as possible, allowing us to drive over it more easily. At least that was the theory. Over the next few hours, the Sun would get higher in the sky and warm the whole ice sheet, melting the surface snow as we progressed further along our route and making it ever harder to drive onwards.

The Argos were relatively slow, with a top speed of around 25kph across the harder, flatter sections of the fast ice. Their diesel engines drowned out any chance of conversation, which, in any case, was difficult because of the cold. The air itself was still and, with our multiple layers and the direct rays from the Sun, entirely comfortable when we had been loading the Argos and standing next to them getting instructions, directions and advice at the start of the journey. Our expedition leader, who would be staying behind with the ship, gave us a quick lesson in how to operate the Argos and then pointed at a mountain of ice around 10km away across the ice sheet. Beyond that, he said, and in almost a straight line, was Cape Denison. We should remember it and, if someone got lost somehow, this landmark could act as a general directional guide for us. He spoke calmly, reassuringly, instilling confidence that we were each capable of completing this journey. By ourselves if it came to the worst. That was his job as leader, to make us believe what we knew, logically, to be impossible.

Once we had started moving, the air rushing past dropped the temperatures we experienced to tens of degrees Celsius below zero. The Argos were completely open vehicles, so we had to hold on to something at all times as we juddered and shook our way across the ice. There was no suspension to speak of on these vehicles, so every bump or crack in the ice jerked us up and down, backwards and forwards almost continuously. We felt every detail of the ice's texture, every drop and rise as the vehicles thudded over the sastrugi – the wave-like ridges caused in the surface of the hard snow by the polar winds – and swerved around the sludgy pools of water.

Even with two pairs of gloves on – merino wool underneath and heavy-duty insulated waterproofs on the outside – my fingers felt so numb that I had to take turns sitting on each hand in an attempt to regain some feeling.

Not every seat was so cold. The Argo I was on, the newer of the two that made up our convoy, had a small windshield and the space next to the driver's seat lay directly behind the engine, which threw out an enormous amount of heat as we trundled along the ice. This was an extreme comfort in this environment and we all took it in turns to sit there, grateful for our hour of relative warmth during the trip.

Our progress over the first hour was good. The landscape changed from the coastline to a pure, flat white and, up ahead, we could see the immense dome of Antarctica's ice cap – the largest single volume of ice on the planet, kilometres deep, sitting so heavily on the continent that it compressed the mountainous ground underneath to below sea level. Closer in, somewhat incongruously, we passed giant icebergs every few kilometres – a reminder that what we were driving across should be open sea. These icebergs, calved from Antarctic glaciers up ahead, would have moved along with the water currents until a few years before. But they were now stranded, locked in place by the thick fast ice that had grown around them. In a few years' time, if and when the fast ice dissipated, these icebergs would be set free and continue on their way into the Southern Ocean.

For our purposes, the icebergs proved to be useful visual waypoints. Once we had lost sight of the shoreline, everything began to look worryingly similar. The dazzling, unbroken white was all there was, with no colour – apart from our own coats and the deep green of the vehicles – or any other features to give any purchase on the scene. The cold made the disorientation worse since it slowed our thinking and made everything blend into one even more quickly than normal. It took concentration in the freezing temperatures to keep our eyes open and stay aware of our surroundings. We had been warned by one of the veteran explorers on the ship that mind maps can get easily scrambled by the incessant white of the frozen plains: people become lost or delirious because every direction looks the same to them, the landscape so vast and empty there is no way for them to get out of their confusion, no leeway for slight error or second chances to fix mistakes.

It was impossible to estimate distances by eye or work out how far we had gone. The icebergs provided targets that we could head towards, to mark out progress. But even when we drove directly towards them, they took a suspiciously long time to reach. However continuously we moved towards them, they just did not seem to get any closer.

After the first 30km of moving across solid ice – as measured by GPS – the snow became increasingly sludgy and our pace slowed. Often we would be reduced to a crawl as we navigated carefully around freshly melted pools of water. Our Argo, the newer one, had tracked wheels and that meant it was less likely to get stuck since it exerted less pressure on the ice underneath. We followed the other, wheeled Argo, in case that one did get stuck and needed to be hauled out. Our attempts to avoid the traps were not always successful, mind you, as we splashed through more than a few pools, sending freezing water all over the unmelted snow and, most of the time, us passengers too.

Three hours into our bone-shaking drive, windswept and freezing, we saw a dark patch of rock up ahead on a high ridge – the first bit of exposed continent we had seen so far on our expedition (every other experience we had had with Antarctica had been of ice, slush or water). This was Cape Hunter, named after John George Hunter, the biologist on Douglas Mawson's 1912 expedition to this part of the frozen continent. It was a big milestone, not only because it banished the more worrying edges of our snow-blindness, but also because passing Cape Hunter meant we were only 13km from Cape Denison, our destination.

The weather so far had been clement, save for the wind as we drove, but we couldn't blame Antarctica for that. As we rounded the bay and on to the last few kilometres into Cape Denison, our vehicles slowed. The wind, however, did not – an icy blast cut through all of my clothes and froze my face instantly.

I had read about the katabatic winds, the geological conditions that caused them to tumble off the polar plateau and the misery reported by Mawson and his men as a result of them. I had seen the pictures of men working around the huts, while standing at impossible 45-degrees angles in hurricane-force winds that threatened to blow them away the whole time. The gusts here are known to reach 250kph, easily more powerful than a tropical storm

Reading about them was one thing. Feeling the winds move across my face, the bone-chilling air dragged across the skin, as if made of steel knives, was another. My eyes streamed and the tears instantly froze to my cheeks. I gripped the rail in front of me on the Argo a little tighter, an unconscious (and completely ineffective) attempt to ready myself against the cold. At 30 knots, though, this wind was just a taster of what this place, on the edge of the Antarctic continent, was capable of. I began to mentally prepare myself for the day we would spend standing in this wind, trying to put out of my mind how it would cut through every fibre of my clothing, through my skin and muscle and, most likely, through to the core of my being.

We were lucky. Within an hour of our arrival, the wind had come to a standstill and Cape Denison turned into an Alpine paradise: the Sun hung, warm, in the clear blue sky, the snow layer around the bay glistened with fresh melt, pools of aquamarine water dotted the bay and, somewhere, we could even hear a running stream. Behind the bay rose a set of rocky hills largely covered in snow and, behind that – way behind that – the dome of the polar ice cap.

Mawson's base is nestled at the bottom of the main valley at Cape Denison. Arriving from the north, we saw a boxy building at the foot of the moraine and, nearby, a few smaller ruins of the smaller sheds where some of the Earth's magnetic measurements were made by his team. The south-facing back of the hut was thick with snow so deep that, when I later approached from that side, the only part of the hut that was visible was the modern version of the pyramid roof, rebuilt by a restoration team in the years before, poking out above the white.

Opposite the front of the main hut, just a few hundred metres away, was the shoreline for Boat Harbour. The day we were there, this was entirely frozen. The ice sheet created as a result of B09B stretched to the horizon and, we knew, beyond.

Mawson's main hut is small: a 7.3m square room, with a small workshop area attached, was home to eighteen men.

Two levels of narrow bunks line the walls on three sides in the main room, with a private cubicle for Mawson opposite the entrance. The rest of the space was a common living area containing a dining table, stores and, in one corner, a darkroom for expedition photographer Frank Hurley. Off to one side was a stove, kept burning day and night, often using seal blubber. Looking after the stove was an important task, not only to ensure the hut stayed warm and people inside could cook food, but also to make sure that any stray fuel did not cause fires in the hut. To that end, every night a different member of the expedition took turns to watch over the stove. Whosever's turn it was to stay up was rewarded with a rare luxury of having a wash – they could warm up some water overnight and give themselves a wipe-over. By the time we arrived, the stove had been cold for almost a century. It stood next to Hurley's darkroom, rusted but recognisable.

In the living area, around where the main dining table would have been, were shelves lined with bottles that once contained sauces, beers or spirits. Piles of cheap paperbacks were stacked against the wall of Mawson's cubicle, evening diversions for some of the hundreds of cold nights the men were here.

At the side of each bunk were the initials of the expedition member who had called it their own, painted onto the wood alongside the years they lived there.

Across the room from the main entrance was a wooden roof beam labelled 'Hyde Park Corner' (named after the park in London), where Xavier Mertz, Cecil Madigan, Francis Bickerton and Belgrave Ninnis had their bunks. Photographs from the time of the expedition show that this area had become known as a kind of social corner for the men, a place where they could go after meals to gossip and smoke their pipes. (The British connection with place names became something of a theme around the base, with one end of Cape Denison known as Land's End and the other, just a few kilometres away, known as John o'Groats.)

Every winter, snow engulfs the huts and, to gain access, conservation teams need to dig through the banks of compacted snow to get to the main door. The huts have been the focus of a conservation programme since 1997 and the interior walls have been carefully worked on, over many years, to remove as much of the ice there as possible. Still, most of the surfaces inside the hut are covered in frost, snow and ice crystals. The floor in the main living area is raised by several feet where a thick layer of ice has built up and any cracks or gaps in the woodwork of the walls and roof are betrayed by large, hard globules of ice. Removing the ice inside the hut is not a straightforward job. These are strong, tough crystals that often require hammers and chisels to make a dent. But the conservationists cannot chip away at the ice with abandon for fear of damaging the wood or original expedition materials that might be embedded within it. It is a careful, delicate process that has taken, and will continue to take, many more years.

When the snow begins to recede around the hut every summer, it reveals a field of rubbish and debris among the rocks surrounding the hut. Mawson's expedition members did not bother to go far when they needed to get rid of unwanted items, throwing out everything here from seal bones to dog chains, wires, old shoes and empty food cans. Two large debris fields, one north and the other east of the main hut, are revealed by the melting snow each summer and, every year, the snow reveals something different to the conservation teams who visit. The debris itself is kept in place to ensure the integrity of the site. Every winter, once again, it freezes over.

Cape Denison was quiet on the day we were there, somewhat contrasting to Mawson's observations of colonies of boisterous, clacking Adélie penguins. Along with emperor penguins, these are the southernmost of the penguins and live on the pack ice around the continent. At Cape Denison, Mawson and his men counted thousands of penguins spread around several rookeries among the hills and near the coastline to Boat Harbour. Though they never made a serious scientific attempt to study them, Mawson's men were nevertheless grateful for the presence of so many birds, which could be a good source of fresh meat whenever the situation arose.

The silence on Cape Denison a century later was due to the looming, unseen presence of iceberg B09B. Since it filled Commonwealth Bay with fast ice, metres deep and locked to the land, the penguins have had a problem feeding. These birds eat krill, part of the plankton in the upper ocean, and the presence of the huge iceberg and fast ice had cut off their ready supply. The nearest open ocean was 70km away and, though the birds do make long marches to and from the sea in order to feed, it is much more difficult for them during breeding season. The birds have a limited range over which they can forage at this time of year – parent penguins cannot stray too far from their rookeries and eggs before they have to come back to relieve their partner at the nest.

Over the hill from Mawson's main hut at Cape Denison, there was a small rookery containing around a hundred or so penguins. Many of them were sitting on or near nests that they had made from piles of rocks. Every time one of them stood up, we could see the gentle curve of an egg under their guano-stained bellies or nestled at their feet. A few hundred metres down the slope from the rookery, the land came to an end. From this shoreline, there should have been open ocean but, instead, all we saw was ice. Beyond the horizon, there was more than 50km of ice.

The penguins ignored us we came near. At that point we could see the effect that the unexpected iceberg had had on these birds. Not only were there fewer of them in total, not only were they quieter, but there also seemed to be a dearth of young. Dozens of eggs in the area in front of us had been abandoned – sitting alone in the rocks or else being incubated without much enthusiasm from the parents, who would stand next to the eggs rather than sit on them or who would just roll the eggs around with their feet. In between the nests were countless dead chicks, penguins that had not survived their first moult.

When food is scarce, adult Adélie penguins tend to give up on rearing their young for that year and prefer to try again the following year when they might have better luck. When an adult has to walk for hundreds of kilometres across fast ice to get to open water from which it can feed, there is little energy left to bring back food for chicks. In that case, the adults just sit and wait. Many of them will die waiting for the ocean to return to Cape Denison and, over the same period, their colonies will not be replenished with new birds. The population will recover, if and when B09B or the fast ice in Commonwealth Bay starts to break up. Until then, these birds are unknowingly fighting a war of attrition with the ice.

The ice sheets and glaciers of Antarctica are built from millions of years of snowfall. Layer upon layer of frozen flakes have landed here over the aeons, never melting because of the frigid temperatures and, as more layers land on top of them, slowly compacting into ice. The snowflakes trap air between them, snatched pieces of atmosphere, each one a perfectly preserved record of the air at the time the snow fell, which then become forever locked into ice sheet. These bubbles of ancient air are our windows into those distant times.

Palaeontologists dig for ancient bones and fossils in the ground and geologists use hammers and chisels to uncover layer upon layer of rock, both looking for ways to understand the deep history of our planet. Palaeoclimatologist scientists do the same by digging deep into ice sheets. But, instead of bones and rocks, they dig out a timeline of ancient climates in the form of long cylinders of ice. At the top is fresh ice, of course, but every metre down is a portal into the past, for hundreds of thousands of years.

The beginnings of this way to see into the past came in 1954 from the Danish geochemist Willi Dansgaard. That year, he published a paper looking at the isotopes of oxygen that were present in batches of fresh water collected from rivers and springs in different climates around the world. His samples included water from the Philippines, Pakistan, India, Argentina, Canada and Greenland, as well as rainwater collected in beer bottles and pots and pans in his back garden in Copenhagen. Water from warmer climes, he found, tended to contain a greater proportion of the heavier oxygen-18 isotope in its water molecules, whereas water from cooler places was made from a greater proportion of oxygen-16. This is a result of how water falls out of clouds as rain or snow, the original source of the majority of the world's fresh water. Though the water molecules in clouds contain both isotopes of oxygen, the molecules made with oxygen-18 are usually the first to precipitate out when the time comes. Only as the clouds cool does the ratio switch back towards the more abundant oxygen-16 water.

In particular, Dansgaard was struck by a sample of ice from Greenland, which contained distinct layers caused by the water melting and refreezing every summer and winter. 'In the opinion of this author,' he wrote, measuring the oxygen isotope ratios in layers of ice would be a way to 'determine climatic changes over a period of time of several hundred years of the past.'

A layer of ice containing a lot of oxygen-16 would be a sign of lower temperatures, whereas ice containing oxygen-18 would be a sign that the environmental temperature at the time period you were looking at was warmer. More than a decade later, Dansgaard worked with a team of scientists and the US Army to dig up an ice core from Greenland. Using his method of analysing oxygen isotopes, he published a paper in the journal *Science* in 1969 that provided a timeline for Arctic temperatures for the previous 100,000 years, the first to use ice cores as a way to study the past.

Scientists still use ice cores to study isotopes and infer temperatures today. But they do much more with these time capsules, too. By releasing the bubbles trapped at different layers of ice, climate scientists can directly measure how much carbon dioxide, nitrous oxide, methane and other greenhouse gases are present, as well as the concentrations of nitrogen and oxygen that existed at different points in the past.

Other factors provide windows into the past, too. Salt crystals are a way to measure the extent of the ocean when the ice was forming. These crystals are transported to the ice surface by wind, from the nearby oceans. Different layers of the ice will have different amounts of salt depending on how far away the ocean was at that point in time.

Dust, ash grains and sulphuric acid trapped in the ice provide convenient date markers – these are evidence of ancient volcanic eruptions whose ejecta would have been blown around the world by the winds. By comparing their positions with records of known eruptions, the dust gives scientists a way to check the age of the layers.

The ice can also be dated using a couple of other methods. Levels of the chemical element beryllium increase whenever the Earth's magnetic field reverses itself and there is a temporary rise in the cosmic rays hitting the surface – geological evidence shows that this happened 780,000 and 900,000 years ago. The ratio of nitrogen to oxygen in the atmosphere changes subtly in a 23,000-year cycle and this can also be tracked in the layers.

Using ice cores to compare the concentration of greenhouse gases in the past with concentrations now has been one of the most important lines of evidence to show that changes in atmospheric concentrations of carbon dioxide and temperature have been connected over several hundreds of thousands of years. Understanding that link began with the analysis of the first ice core dug up at the Russian station Vostok, in the remote East Antarctic, in 1985, which gave scientists an insight into the past 150,000 years of temperatures there.

Most modern ice records come from ice cores drilled out of Greenland and Antarctica. Whereas the ice flows more dynamically in Greenland, Antarctica has been covered in ice for more than 30 million years – the oldest continuous records only extend back 123,000 years for Greenland cores but go back 800,000 years for Antarctica. Given that our direct measurements of greenhouse gases such as carbon dioxide only go back to the 1950s, ice cores are a way to look into the deep past. (We can check that ice cores are a faithful record of past atmospheres by looking at overlaps – in some parts of Antarctica, such as Law Dome, the snowfall is so high that the air from the 1980s is already encased in ice.)

For a long time, the flagship ice record was a core drilled to 3.6km deep in the mid-1990s at the Russian Vostok station. That contained a climate record that went back 400,000 years and spanned four ice ages. Among many successes, that core provided unequivocal evidence (supported by earlier cores dug from the same location) that carbon dioxide levels in the atmosphere have risen in step with temperature for many thousands of years.

Analyses of ice cores by the British Antarctic Survey show that carbon dioxide levels in the atmosphere were stable over the past millennium and then started to rise in the nineteenth century around the time of the Industrial Revolution. The level continued to rise so that now it is 28 per cent higher than at any point measured in the ice cores. Methane has also shot up in the past two centuries to more than 134 per cent greater than prehistoric highs.

Both the magnitude and rate of increase of these greenhouse gases has been unprecedented over the past 800,000 years. The fastest natural increase in carbon dioxide according to ice core data is around 20 parts per million (ppm) by volume in the atmosphere over the course of 1,000 years, around the end of the last ice age 12,000 years ago. At the turn of the twenty-first century, that same increase happened in ten years.

In 2004, the Vostok core was superseded by an ice core 3.2km deep, which held climate records back to 800,000 years, drilled at Dome Concordia in the East Antarctic, 500km from Vostok. Though not as deep as some previous cores drilled in the area, its bottom sections were so well preserved that they gave scientists a much more robust way to look even further back in time.

Dome C, as the location is often abbreviated, is one of the snowy summits on the Antarctic ice cap where temperatures can drop to -50°C and scientists drilled the thick white ice there as part of the European Project for Ice Coring in Antarctica (EPICA). At its oldest, parts of this ice core came from so deep within the ice sheet that the bubbles of air had been squeezed out completely and the frozen water had become as clear as glass.

Collecting the core so that it maintains its integrity and scientific value is not trivial. The scientists used a drill attached to a 10cm-wide, 3.5m-long tube that collected several metres of ice in one go, which was then hauled up to be removed and stored. Aside from the blizzards and numbing temperatures, the ice itself posed challenges as the scientists drilled down. The ice from the first 1,000m, for example, is relatively brittle because the air bubbles are trapped in between the ice crystals and bringing it up to the surface could cause it to shatter. The drill itself can also become jammed by the small chunks of ice that are thrown off as it grinds its way through the ice sheet – more than once the Epica team lost drills in the ice like this and had to start again with new equipment. The losses made the drilling team wary of the conditions and, afterwards, they tended to drill shorter lengths before bringing the cores back to the surface. In turn, that increased how long the whole process took – at 3km deep it took an hour to bring a sample back to the surface.

The desire to go further in time, to build a more comprehensive climate record, goes on. Having almost finished analysing the Epica ice cores, scientists are already looking for ways to get an ice core that goes up to 1.5 million years into the past. A likely location for this core is Dome Argus (Dome A), the highest of the snowy summits on the Antarctic ice cap, 4,000m at its peak.

Dome A is one of the least accessible parts of the continent. It was the last of Antarctica's peaks to be conquered, finally scaled in 2007 by a team of Chinese explorers after a month-long 1,300km trek across the ice from the Zhongshan base on the Antarctic coast. The ice is thick here but it gets very little snow – up to 1.5cm of water-equivalent every year compared to 3cm at Dome C and 50cm for some of the coastal sites on the continent where cores are sometimes dug. Every metre of ice from Dome A would contain a lot more scientific value than the cores dug elsewhere. The Chinese began building their Kunlun base at Dome A in 2009 after the International Polar Year, their third on the continent, with the aim of finding sites that would be suitable for drilling a new ice core.

Back in the lab, once the ice cores have been sectioned into salami-thin slices, crushed up to release the air from their bubbles, had their ice crystals measured and the dust extracted, the ancient ice is left to melt. The water drains into the sink in the climate lab in which it ended up and, thousands of miles from where it fell as snow and hundreds of thousands of years after it first got trapped on Antarctica, the water enters the Earth's hydrological cycle once more.

According to the Intergovernmental Panel on Climate Change (IPCC), the Arctic has been warming since the 1980s at around double the global rate. The sea ice there has declined by around 13 per cent per decade and the region is predicted to have ice-free summers by the middle of the twenty-first century. The permafrost in the surrounding areas has warmed by 0.5 to 2°C since the late 1970s.

At the other end of the world, the picture is a little more mixed. The strongest rates of atmospheric warming are happening in the Western Antarctic Peninsula (at around 0.2 and 0.3°C per decade) and also the islands of the Scotia Arc, which have also seen warmer oceans and a reduction in the amount and duration of winter sea ice. Milder warming has happened at other edges of the continent, near the Bellingshausen Sea, Prydz Bay and the Ross Sea, but some of the areas in between have cooled.

In the past three decades, there has been an unusual set of collapses in the ice shelves in the West Antarctic Peninsula. This region of the frozen continent is out towards South America and is surrounded by warmer waters anyway but the peninsula here has warmed by 2.5°C since 1950, making it one of the fastest-warming places in the world. Warmer air tends to melt the surface of the ice shelf and create pools of meltwater, which can trickle down and widen cracks in the ice. Less sea ice in the region around the ice shelves (due to warmer oceans) also means that the ice shelves are subject to rougher seas and bigger waves, which can smash against them and flex the ice, creating stresses and weakness.

In 2008, the Wilkins Ice Shelf in the Western Antarctic retreated by 400 square kilometres. The Larsen Ice Shelf disintegrated in spectacular fashion in 2002: in just over a month from February of that year, 3,250 square kilometres of ice disintegrated. Pools of meltwater were seen on the surface of the ice shelf by satellites in the months beforehand and the final result was a melange of huge icebergs and boulders of ice – some 500 billion tonnes in all, according to estimates – that eventually floated off into the Southern Ocean.

Whether or not you think any of this is a bad thing depends on your specific point of view. Environmentally speaking, the warming and reduced volume of ice will have big impacts on the wildlife in the area. On the other hand, the warmer waters will open up new shipping routes during the summer and expose new areas for mineral and oil mining.

Less sea ice means more light entering the upper part of the oceans, with implications for the food webs in cold waters. For example, the bottom of the chain, so-called primary production, will switch increasingly to water-borne phytoplankton rather than ice algae. Animals that are evolved to live in this part of the world, from fish to polar bears, will have to adapt to the new conditions and sources of food, or die.

Sea ice itself is an important habitat for many species. Krill, a basic food source for many higher animals in the polar marine ecosystems, use sea ice as a place to live and also get protection from predators. Seals, walruses, polar bears, penguins and other birds use sea ice as makeshift dining tables during their hunting, a place to set down and eat their food. As important, if the distribution of sea ice changes, is the fact that the overlap between areas where hunters and prey live will shrink. Some species of seals, such as bearded seals in the Arctic and crabeater and leopard seals in the Antarctic, give birth on pack ice.

Antarctica is no place for people. It is a hard environment, unforgiving and antithetical to the idea of life. Anyone who chooses to live or work here absorbs an automatic set of rules for the road, ideas that they know will allow them to eke out an extra margin of safety in this tough, changeable environment. Keeping in mind which direction the wind is blowing, relative to the way the ship needs to move, relative to the floating icy obstacles in its path, becomes a ritual. Carefully applying woollen underwear, gloves, hats, scarves, down jackets and water and wind protection layers before walking through any of the external doors onto deck moves into the subconscious. Nothing – people or ships – moves too fast; everything is calculated, plotted, backed up and rechecked.

This continent is not just cold, windy and wet. It is the extreme of all of those things. Leave a hole in your personal armour – a glove left untucked into a sleeve or a gap around your neck where you forgot to put on a scarf and thought it would be too tedious to take off your multiple layers and rectify the situation properly – and the frozen winds will find and punish your laziness as soon as you get outside. Just a few minutes of exposure to the frigid, desiccated air can leave fingers numb and useless. The cold begins as a stabbing sensation, then it becomes searing and, in the moments around when you lose feeling completely, it sends your nerves into a symphony of confusion.

You discover these things when you make mistakes. At one point, I took a glove off to type an email while my computer was connected to the satellite internet connection at the top of the ship. I was impatient to respond to something immediately instead of following the sensible protocol of only ever uploading and downloading emails while outside, leaving the reading and writing of them for the inside of the ship, where the conditions were more normal. My fingers soon stiffened as I typed, turned red, then white, and as they seemingly disappeared from my hand, they felt bizarrely hot. Painful, boiling hot, as though I had plunged them into freshly made black coffee.

On Christmas Eve, a blizzard hit our ship and 50-knot winds – relatively mild for Commonwealth Bay – made it difficult to stand up straight on the deck. Anything not tied down outside slid and tumbled across the metal. The swirling snow enveloped our ship in a cloud of blinding grey-white and the only information we had about the world outside our immediate vicinity came in the form of the howls and whines of speeding air.

Weather can change completely in a matter of hours around Antarctica. In this great wilderness, wind, snow, ice and temperature are not just another layer of information about your environment, something that you need to take a passing interest in as you make your way from one place to another. These things are your environment. On the frozen continent, no one makes concrete plans, they only hold in their heads possibilities that depend on the weather. People here can't work to fixed timetables and the only thing they know for certain is that they'll have to adapt and change their intentions every few hours.

This takes some getting used to, especially if you normally live in a modern city. I live in London and travel to a lot of places across the UK and around the world. All of this is done within the vast global infrastructure of airlines, hotels, money, telephones and websites. All of these have backups and alternatives that allow individuals to get around blockages and other problems without requiring any expertise. I want to go to work and the London Underground line I want is broken? I can look into getting a bus. Or a taxi, if it is urgent. If none of those are on the street, I could call a car on the phone. If the worst came to the worst, I could walk. All of this extra hassle would be unwelcome and, perhaps, expensive, but I would get to where I want to go. If there were a delay in reaching where I needed to be, it might be for a few hours. Anything over four hours (the exact number depends on who you are and what you're missing as a result of the tardiness) would be a catastrophic delay, something that would raise hackles and might prompt letters of complaint. In any case, during any city-type delay, you'd be able to keep everyone updated with what you were doing, where you were, tap out messages and perhaps even have useful work conversations on your data-enabled phone.

None of that infrastructure is available in Commonwealth Bay. We were on our own here. Sure, the authorities knew where we were and other local ships might be able to find our general location on a map, if needs be. We were hooked up to satellite comms and GPS. But the closest people were hundreds, thousands of miles away. In rough seas, they were days of travel away. Helicopters and planes could not reach us – too far from base for the former and nowhere to land for both former and latter.

What constitutes a catastrophic delay here? It's impossible to tell. You might wake early and want to go from one place to another by the afternoon, a modest distance in what looks like a still, sunny morning. It might even happen. But the continent teaches you, eventually, to accept that this dream, however modest, might not happen today. It might not happen tomorrow, either. Perhaps not even next week. Antarctica laughs at tight timetables.

This continent keeps you on your toes. With such raw beauty all around, it is easy to get lost in wonder, to forget the layer of danger that lies beneath every square centimetre of snow and ice. Treat this place as common, easily understood or normal, lose your concentration or forget that this seemingly inanimate environment can easily trap you without even noticing, and you're asking for trouble. Antarctica is not a place that forgives mistakes easily.

Blizzards can obscure landscapes but the continent

shifts in other ways, too. An ice floe or ice sheet might look stable until you put your foot or a vehicle through it. The wind might feel benevolent until it picks up, starts to swirl and screams past your ear or face, stripping it of warmth, moisture and function. The Sun might be low or dim in the sky until you realise that its levels of UV radiation are sky high, thanks to the relatively thin air on the continent and the angle of our star in the sky. A snowfield might look pristine until you fall through a crevasse half-way across it.

One of the scientists told me of the mixture of exhilaration and fear that comes with climbing mountains in Antarctica. Climbing through deep snow, tied to a colleague behind him, he focused on every noise that his boots made into the soft surface. One foot in front of the other at a steady, careful pace. With each step, the snow crunched as it settled beneath his boot. His heart pounded, adrenaline pumped. Because every crunch also brought the possibility that it was the start of a catastrophic fall, the break of a snow bridge covering a crack in the ice. With every step he readied himself to throw his body forward if the worst happened. The imagination, he said, runs wild in this sort of environment, with that much stress layered onto it.

People use technology where they can to keep oriented and prevent accidents but, in reality, even the finest developments of the twenty-first century cannot mitigate against the savagery of the ice and the effects it can have on us. Thus we become each others' eyes and ears. The first sign that someone is succumbing to the conditions is when they stop reacting to others, turn inwards and go quiet. As careful and alert as we were told to be towards the environment whenever we went outside for extended periods of time, we were told to keep an eye on each other just as much.

Knowing where people are is also important in preventing yourself from losing your bearings. There are few visual clues that you can keep in your head to prevent getting lost. Everything really is the same colour and the lack of contrast can be destabilising. Because there are no plants, there is no pattern of green in any direction, nor the browns or greys of soil and rocks. A day on the ice is all it takes for senses to become heightened to the different shades of white, a reaction to the lack of normal stimulation, but ice sheets are still remarkably easy places in which to get lost.

Not only is everything white but one iceberg can also look deceptively like another when you're distracted. Distances are also supremely difficult to judge and time seems to move oddly in the unchanging landscape. Driving back from Cape Denison after our visit to Mawson's huts, I could see the ship on the horizon ahead after several hours, a faint dot with nothing but a flat ice sheet between it and us. Though we moved fast across the ice, the ship just never seemed to get any closer. Every time I looked, tired and cold, at the dot, it was still the same.

As such, mind maps can get scrambled within seconds. You thought you knew which direction you were headed in half an hour ago? Get distracted by cold, hunger or tiredness, and you're moving in the wrong direction, perhaps even backwards. Thank goodness, then, for the objective measure of GPS to keep things moving in the right direction and to measure distances without any need for your compromised, subjective input. If the batteries die, however, you're on your own. And if your vehicle runs out of fuel, freezes or refuses to restart after it has been pulled out of a snow flurry, you're in trouble. The nearest person is at least several hundred miles away, perhaps a thousand. This is not a place replete with second chances – you can't simply ask for directions or wait for someone to wander past to get help. If you get lost, if no one else knows where you went, the chances of anyone finding you in the vast icy tundra are not high. For that reason, anyone going out onto the ice, out of sight of base or ship, takes along their own second chances – backup GPS devices, protocols that require direct contact with base every few hours, food and shelter if they are forced to camp out for a night. And many, many spare batteries, kept warm against the body to stop them running out of juice.

Antarctic veterans will tell you to be mindful of these dangers, learned from several seasons on the ice, as the ground rules for any trip across the empty landscape. They'll tell you to treat the landscape with respect, something you can perhaps get away with not doing back home in the civilised, conquered world.

For them, this exquisitely detailed awareness of the Antarctic environment is not a chore, though it is a sort of entry price to the privilege of experiencing a continent that gives you indescribable silence, epic skies and a sense of isolation unmatched anywhere else in the world. In fact, the need for awareness is the privilege itself: this is an environment so alien and dynamic that it demands attention; it requires you to understand it before you can exist on it.

But, despite all these things that would make creatures wither and die, despite all the things the continent conspires to do to prevent it, life does cling on here. Algae cling to the underside of the sea ice; lichen grow in tiny patches across the surface of high-altitude rocks; penguins, petrels and seals patrol the coastlines, persisting along the slimmest edges of liveable temperature, water and nutrients. Their fortunes are tied entirely to the jolts of their environment, the combined movements of ice, water, wind and sunshine.

In fact, it can be downright surprising how far life will go to exist in this stark, harsh place.

In January 2013, towards the end of that year's summer season in Antarctica, John Priscu trudged across the snow to a metal shipping container where he had set up his laboratory on the edge of the Ross Ice Shelf, around 650km

from the South Pole. In his arms he held a sealed, grey
metre-long tube, containing a precious sample of water
drawn from Lake Whillans, a shallow body of water that lay
beneath 800m of Antarctic ice. The lake had been isolated
from the rest of the world for at least 120,000 (and possibly
up to a million) years. Priscu had come to find out if this
frigid, lonely, dark place contained any life.

It had taken him six years of meticulous planning and
$20 million to get his sample, the first ever from a subglacial
lake in Antarctica. His team not only had to bring hundreds
of tonnes of engineering equipment to one of the most
remote parts of the world, they had to operate in temperatures of -15°C while drilling a 60cm-wide borehole through
the ice cap, all the time enacting their strict protocols to
ensure that whatever they collected did not become contaminated by the tools they had brought.

They collected 30 litres of water over the next four days,
flew the refrigerated samples back to Priscu's lab in Montana
State University and spent the next year analysing what they
had found – extracting cells, culturing them, sequencing any
DNA. In August 2014, they reported in the journal *Nature*
that they had found 130,000 cells in each millilitre of the
water, a density of life similar to that in the deep oceans,
and there were almost 4,000 species of bacteria and archaea
in all. 'I was surprised by how rich the ecosystem was,' Priscu
told *Nature*. 'It's pretty amazing.'

The extent to which life can survive on Earth can often
be surprising even to those who go looking in the most
extreme places. Around the Antarctic continent's coastline,
within the edges of the Southern Ocean, life is abundant on
the shorelines and even in the deepest, coldest water. On
the wasteland of the Antarctic plateau, though, one of the
driest places on Earth, no creatures or plants have a hope.
A few species of birds swoop through the cutting winds
above the surface but they don't nest on the ice cap itself.

No ecosystem can be possible, surely, in a place so devoid of warmth, moisture, nutrients and, to human eyes at least, even a crumb of comfort.

Underneath the ice, though, it is a different story. Scientists began to think about looking for life there in the early 1990s, when ice-penetrating radar showed the first evidence of subglacial lakes. These lakes form under the ice sheets and are fed by water melted by the Earth's geothermal heat. These are ancient lakes, formed when the melt water is forced by gravity and ice pressure to flow into the hollows and valleys of the Antarctic landmass under the ice. Lake Whillans is one of almost 400 subglacial lakes that have been found so far under the Antarctic ice using gravity and seismic mapping. It is 60 square kilometres in area but only two metres deep.

Priscu's team sterilised all their equipment with hydrogen peroxide and ultraviolet radiation and spent seven days using a jet of boiling-hot filtered water to penetrate the 800m of ice down to the top of the lake. The water they pulled up was a golden liquid, rich in minerals and biological cells, the first of which were identified under a microscope just hours after samples were brought into the lab near the encampment.

The life in Lake Whillans has no access to sunlight, so it needs other ways to power metabolism. Many of the archaea in the sample seemed to use the energy contained in the chemical bonds of the ammonium present in the lake, in order to fix carbon dioxide and make the materials they needed to live. Other organisms used the energy and carbon in the methane in the lake to stay alive. These raw ingredients most likely came from the breakdown of organic materials that had been deposited there hundreds of thousands of years earlier when the Antarctic was warmer and the sea inundated the area around Lake Whillans and the West Antarctic more generally.

A few thousand kilometres away, Russian scientists have been working on a similar project to sample ancient water for more than twenty years, from a vastly bigger subglacial body of water that is thought to have been isolated for anything up to 15 million years – Lake Vostok. Covered in 4.2km of ice in Antarctica's eastern ice sheet, Lake Vostok is the continent's biggest subglacial lake, covering an area of around 14,000 square kilometres, 670m deep and sitting underneath the Russian research station that bears the same name. On a hostile continent, the area around Vostok base is a particularly unpleasant place – temperatures here have been known to drop to -89°C, the coldest ever recorded on Earth.

In 2000, scientists found microbes embedded within the bottom 100m of an ice core that had been drilled to almost 4,000m deep at Vostok station and this spurred them on to continue further into the lake, to work out what lay inside. There were plenty of stops and starts as scientists argued about the best way to reach the lake without contaminating it. After more than fourteen years of metre-by-metre work during Antarctic summers, Russian scientists finally broke through to the lake in February 2012, right at the end of that year's season before the temperatures in the area dropped too low for them to continue working safely. The team retrieved a sample of the water, probably from a pocket of melt water just above the lake, and found evidence of microbe DNA.

How can anything live in such extreme conditions, embedded within ice that is several degrees below 0°C and at pressures of more than 400 atmospheres? One idea is that they live within a network of liquid-filled veins that exist in the glacier between the ice crystals, where salts accumulate and prevent freezing. Organic acids could also collect in these veins and the chemical reactions between them might be enough to provide energy for the bacteria there, which are starved of sunlight or any other warmth.

None of this work is easy. A few months before Priscu took samples from Lake Whillans, a British team led by Martin Siegert of the University of Edinburgh tried to use similar hot-water-drilling technology to sample the water from Lake Ellsworth, a body of water that is 12km by 3km and around 150m deep and sits under 3.4km of ice in the West Antarctic ice sheet.

Like Priscu's Lake Whillans project, Sigert's team planned to use a jet of hot water to drill through the ice. A decade in the planning, engineering teams had installed the heavy equipment in December 2011 and all of the equipment for use in the borehole had been pre-sterilised in clean rooms in the UK and was only unpacked when in position in the Antarctic. The samples of water were to be collected in pressurised titanium cylinders, to preserve their contents, and only unsealed back in the lab in the UK.

In November 2012, the team arrived on site and began drilling, using the equipment that had been placed the previous year. Temperatures outside dropped to -25°C on days when there was no wind and visibility dropped to a few metres whenever there were blizzards.

After a month of drilling, however, things were not going to plan and in the early hours of Christmas Day, Siegert called a halt to the project. Because of the large amount of hot water needed to drill down to 3km, Sigert's technique involved first melting a reservoir of water 300m below the surface and then drilling the main borehole through that reservoir and on to their target below. Though they managed to create their reservoir and keep it open for forty hours, they couldn't locate it with their main borehole, even though they only drilled that two metres from their pilot borehole. After their attempt failed, Siegert said that his team would be back to try again, but any return will take several years to plan.

Before either the British or American teams had begun

drilling, Priscu, a thirty-year veteran of polar research, said that one of the most important results of finding life under the ice would be how the world viewed Antarctica. 'Instead of being a big block of ice, we're going to look at it as part of the living ecosystem on Earth, which has never really been the view,' he said. 'The surface of Antarctica is pretty harsh. We see very little to no life at the surface – it's too dry, too cold. Once you get down to the bottom, it's warmer, there's liquid water, it's a much more clement environment down there. You just have to have a special set of organisms to know how to use it.'

The knowledge that life can exist in such strange places will be useful in another existential question that has bothered humans ever since we started to look up at the stars and wonder if there was life somewhere out there. Subglacial lakes such as Vostok, Whillans and Ellsworth are the best Earth-based analogues we have for the environments that might exist on Europa and Enceladus, moons of the gas giant planets Jupiter and Saturn respectively. Multiple lines of evidence from probes and Earth-based observations show that these worlds are covered in thick layers of ice, with liquid water underneath. Mars is known to have ice sheets at its poles and also volcanic activity – if the two are combined there might be liquid water under the ice somewhere.

If life can exist thousands of metres under the ice at Vostok and other subglacial lakes, that raises the chances that it might also survive in the depths of alien environments in deep space.

The Antarctic is just one place where life seems to cling on, spurred on by the presence of a little liquid water despite otherwise inhospitable conditions. The continent is a desert by definition because it is a place that receives little or no rain. More familiar hot deserts are also difficult for life, of course. Without rain or plants, there is little hope of an ecosystem. Anyone who has experienced the searing heat and burning sands of the Sahara or the numbing cold of the Antarctic plains will tell you that these are desolate, lifeless places, other-worldly and devoid of the familiarity of things growing and moving.

Even by any of these definitions, the Atacama Desert in northern Chile is special: the oldest and driest desert on Earth, in some parts it hasn't rained for hundreds of years. This high-altitude region, between the Andes to the east and the Coastal Range to the west, is covered in boulders and a desiccated, dark red soil that is so alien to human senses, it has been used by various space agencies as an analogue to Mars to test the robotic landers they want to send to explore the red planet.

For a long time, it was assumed that the Atacama was lifeless, a dead place on a planet otherwise teeming with life. Without any liquid water, how could any form of life survive here, save for the few cacti and other spiky plants that dot the terrain? But survive it does. You could even argue that it thrives. At the El Ratio geothermal field, located at 4,300m above sea level in the Andes, cyanobacteria survive in the harsh radiation environment by building tiny casings made from silica rock. Cyanobacteria also live a few millimetres underneath the surface of salt rocks in the Yungay region, occupying space between the crystals of sodium chloride. There is no liquid water in their surrounding environment and their existence depends on the condensation of water within tiny pores in the rock. At 6,000m high, close to the summit of the Socompa Volcano, are mats of acidobacteria. Nothing should live in the barren soils of the Atacama, but living things have found a way.

As we begin our search for life among the stars, it is worth taking stock of the limits of life on our own planet. In recent decades, we have come to realise that, no matter what the physical conditions, wherever there is a hint of liquid water on Earth, there is life. Whereas we might once have thought of certain temperatures, pressures or chemical conditions to be antithetical to life, we've come to realise that we've unconsciously meant life comfortable for humans and other complex animals. Every niche on Earth is active with life, whether in freezing-cold lakes in Antarctica or boiling-hot underwater vents in the darkness of the ocean trenches, within pools of strong acid or kilometres underground in mine shafts. By these measures, the Atacama's Martian-like soil is nothing so strange or hostile at all. In 1974, NASA exobiologist Robert MacElroy proposed the name 'extremophiles' for this strange class of organism.

There are plenty of extreme environments on Earth, and all of them are colonised in some way by life. Hot springs and geysers that have erupted around volcanic formations on the surface; salt flats and lakes super-saturated with salt (for example, the Great Salt Lake in Utah); deserts where there has been little or no rain for years, such as the Atacama in Chile. The deep sea contains not only almost-freezing temperatures, but also hydrothermal vents on the sea bed where water can get as high as 400°C but remain liquid due to the intense pressure at depth. Even in the coldest, driest place on Earth where it is dark for half the year – the dry valleys of Antarctica – communities of cyanobacteria, lichens and green algae have adapted to the lack of light and occasional dustings of snow that, when they melt, produce a small amount of water for a brief time.

All of these environments are what we would call extreme – well outside our own comfort zones of temperature, pressure, acidity, aridity or radiation – and the organisms that have evolved to live in them are extremophiles. The vast majority of these organisms are microbes, single-celled organisms in the domain of archaea and bacteria, but some extreme environments are known to support more complex eukaryotes too, including some vertebrates. Eukaryotes are the branch of life that eventually led to multi-cellular organisms, including plants, dinosaurs, elephants and us.

The key point is that all of these organisms use the same biochemistry as humans: DNA for genetics, amino acids in proteins and the same basic metabolic reactions. All of these bizarre organisms are related to us and every other form of life we know about, all of us descended from a common ancestor many billions of years ago. Our standard biochemistry, in other words, is versatile enough to survive in the toughest of places.

Hyperthermophiles, for example, can survive in temperatures greater than 80°C, which far exceed the tolerance levels of normal organisms. The highest temperature ever recorded is for an archaea called Geogemma barossii (nicknamed 'Strain 121') found in a water sample from an active black smoker hydrothermal vent called Finn, located in the Mothra hydrothermal vent field in the north-east Pacific Ocean. Around a thousandth of a millimetre in diameter, it has a tail-like projection known as a flagellum to help it move around and has been shown to grow in temperatures up to 121°C. Before that, the record was held by Pyrolobus fumarii, which can grow at temperatures up to 113°C. Cultures of Strain 121 kept at 130°C for two hours did not die – in fact they managed to grow again when their temperature was reduced back to around 100°C. (The discovery of microbes that can exist at such high temperatures might worry those who want to, say, sterilise hospital equipment, which is generally autoclaved to around 120°C to keep it free of germs. Fortunately for us, Strain 121 finds it difficult to grow at human body temperatures.)

Lynn Rothschild, an astrobiologist at the NASA Ames Research Center in Moffett Field, who studies life on Earth in a bid to understand its limits and the possibility that it exists somewhere else, listed cyanobacteria, Bacillus,

Clostridium, Thiobacillus, Desulfotomaculum, Thermus, lactic acid bacteria, Actinomycetes, Spirochetes and many others, in the list of thermophiles for a review in *Nature*. Not to mention archaea including Pyrococcus, Thermococcus, Thermoplasma, Sulfolobus and the Methanogens. The enzymes in some of these hyperthermophiles, such as amylopullulanase, can keep working at temperatures of up to 142°C.

The upper limit for eukaryotes, in contrast, is around 60°C. Some protozoa, algae and fungi can survive at those temperatures. Mosses and vascular plants can survive up to about 50°C and fish can go to 40°C. That's also about the limit for comfortable human existence (our body temperature is maintained at 37°C).

The problem for most organisms is that high temperatures tend to destroy their biomolecules. Most proteins will irreversibly denature and stop working above a certain point (think of an egg white becoming solid when you boil it), DNA is destroyed and the covalent and other bonding required for cells to keep working will start to go wrong when there is too much energy in the system. For plants, an important limit is at 75°C, when chlorophyll degrades and photosynthesis will therefore stop.

At the other end of the temperature scale, low temperatures can slow down cell function to a level where the organism cannot survive. And when water freezes and expands into ice, the crystal can tear apart cell membranes. Even so, lots of things can survive in the deep cold. Biological cells – including human eggs, sperm and tissue cultures – can be preserved indefinitely if they are flash-frozen in liquid nitrogen (at -196°C) and the water cannot freeze into big crystals. They stop working at this temperature, though, and need to be warmed back up to ambient temperatures before they come back to any semblance of life. The Himalayan midge, on the other hand, can keep functioning to -18°C.

Microbes have been measured at -20°C in ice and the Antarctic cryptoendolithic (which translates as 'living hidden in the rock') lichens can carry out photosynthesis at -20°C. Water can, of course, remain liquid at even lower temperatures if it is mixed with other things such as salts or organic solvents. Mixtures of methanol, ethylene glycol and water can sustain enzyme activity down to -100°C. Given the right mixtures and catalysts in solution, it is possible that there is no low-temperature limit for life.

Radiation – either particles or electromagnetic waves – can damage DNA in high doses. The Earth's atmosphere protects us from the lethal doses that pour out of the Sun every day so there is little need for life forms to adapt to it. But there are artificial situations where radiation levels are high, for instance around nuclear reactors or temporarily around medical scanners, and there are organisms that can live within radiation regimes that nothing else could tolerate for very long without dying or becoming seriously damaged. The bacterium Deinococcus radiodurans, for example, is well known for its ability to withstand immense amounts of ionising UV and gamma radiation without suffering fatal damage to its DNA, most likely by having souped-up DNA-repair mechanisms inside its cells. Other organisms that have been shown to withstand extreme radiation include Rubrobacter radiotolerans and Rubrobacter xylanophilus, isolated from hot springs in Japan, Portugal and the United Kingdom, and the green alga Dunaliella bardawil, which manages to tolerate high radiation by accumulating beta-carotene into its cells.

Piezophilic species (sometimes called barophiles) grow at extreme pressures of up to 700–800 atmospheres and exist in the dark areas deep in the ocean. The most extreme environments where these organisms, such as the large amoeba-like protists called Xenophyophores, have been found include Challenger Deep at the bottom of the Mariana

Trench in the Pacific Ocean, 11km below the surface of the sea.

Life finds ways to thrive in even the most extreme of environments. Their existence and environments might be alien to us but, in all cases, these extremophiles rely on water. Indeed, so ubiquitous is the importance of water for the functioning of life that its presence has become a shorthand for life itself: where there is water, so there is life. It is this trait, as we will see, which has allowed us to go looking for traces of life beyond the confines of our own world.

Part IV

———∽∽∽———

SPACE

NINE

The Moon

The Earth isn't the only place in the solar system that hosts water. Some of the moons of Jupiter and Saturn, including Europa and Enceladus, are giant balls of water covered in thick layers of ice. There is evidence for water in the form of hydroxyls (a sub-unit of the water molecule, made of a pair of chemically bound oxygen and hydrogen atoms with an overall negative charge and which is usually found bound to other elements) in the soil of the Moon and large amounts of ice at the poles.

Mars once had liquid water on its surface, though the low temperature and atmospheric pressure now means that it is not there any more because it would quickly freeze and then evaporate into the atmosphere. There are still huge glaciers at the bottom of craters on the red planet, though, and the polar caps are covered in ice. Water did once flow on the surface, which we know from pictures of channels and other geological features that could only have been produced if water had routinely passed over them at some point in their history. Further evidence came when NASA's latest Mars lander, Curiosity, took pictures of its landing site only to reveal that it was on an ancient river bed in the Gale

Crater, at the foot of its intended target of Mount Sharp.

Lumps of water ice are known to circle Saturn as part of its majestic rings. On one of its moons, Titan, there are underground oceans containing water that could fill the Earth's oceans twenty-nine times over. Jupiter's moon, Ganymede, has the equivalent of thirty-six Earths' worth of water and Callisto has twenty-seven Earths' worth. The insides of Uranus and Neptune are suspected to be gargantuan balls of ice and even Mercury, a planet with a surface so hot that it would melt metal, has water ice deep in craters on its poles.

Other moons of our solar system also contain water. In fact, apart from Jupiter's volcanic moon, Io, all of the major satellites of the gas giants seem to contain varying amounts of solid (and possibly liquid) water, depending on how far they are from the Sun and also their parent planet. Titan, Europa and Enceladus are the ones that space scientists are most interested in but Callisto, Ganymede, Triton and many others also contain significant proportions of water. Even the loose rubble beyond Neptune, the Trans-Neptunian Objects, contain comets and ice-filled rocks that are several hundreds of kilometres across, such as Quaoar and Haumea.

This confirmation of water outside the Earth came gradually over the second half of the twentieth century. From a point when scientists thought Earth was the only place with water, it is now confirmed that almost every region of the solar system has it (or had it) and a great deal more than previously thought had liquid water on their surface.

NASA's Mariner 9 first saw unmistakable evidence for a watery Mars from orbit, in 1971, when it sent back pictures of channels on the surface of the planet that looked like dried-up river beds, and Viking confirmed these findings. In the late 1980s, Voyager flew past Jupiter's moon, Europa, and found hints of a liquid water ocean underneath the ice. (Arthur C. Clarke even wrote about it in the novel *2010*, his sequel to *2001: A Space Odyssey*. After a probe spotted what looked like chlorophyll in the moon's sea, the mysterious monolith warned the nearby Earthlings to stay away from the moon and attempt no landings there.) In 2005, the Cassini probe flew into plumes coming out of the bottom of Jupiter's moon Enceladus, finding that they were huge geysers of water.

'It's certainly been interesting to see these new worlds come into the picture – Enceladus in particular. We were happily going along thinking about Mars and Europa and then, "Kapow! Hey, look at this world!" That's exciting and motivating, there's something new,' says NASA astrobiologist Chris McKay. 'I'd love to see another one but I don't see much chance for much else in our solar system. But then I wouldn't have predicted Enceladus either.'

Astronomers are finding an ever-greater number of planets outside our solar system and, when the technology gets good enough in the coming years, they will be able to look at which chemicals are present on these worlds. You can expect water to be at the top of their search list. Many expect it to be present in a lot (if not all) of the planets they find orbiting other stars.

This grand search for water on other worlds is, in shorthand, a search for life. Since we don't have a method to directly detect life – a foolproof biological scanner of some sort that we can take to the surface of planets and confirm, yes, there is or is not a living thing present somewhere – we look for water instead. NASA's stated aim is to 'follow the water' because all life on Earth, every single organism we've ever seen, requires liquid water in order to grow and reproduce, for its internal structures and membranes and biochemistry to work.

'The chemistry of water, the abundance of water, the various physical properties of water are all interesting – it's not a bad thing to start with,' says Lynn Rothschild, an astrobiologist at NASA's Ames Research Center in Moffett Field.

'Organisms can be dormant in a very dry state and some organisms can survive for very long periods of time profoundly dry, but they don't grow or reproduce in that state,' says McKay. 'In order to grow or reproduce, every organism requires liquid water. There was a hope once that we could find organisms that could grow on ice, that would literally be frozen and metabolise, but this is just not the case. There are organisms that grow in ice, at low temperatures, but they grow by having water still be liquid either due to thin films or salts keeping the solutions liquid. Liquid water is observed to be a common requirement for growth and reproduction for life on Earth. The strongest argument for water is just this observation. It's purely empirical. Everybody needs it on Earth; we're going to search for other environments that have water; we know water works.'

Looking for liquid water means focusing on planets that are in the habitable zones around a star – the so-called 'Goldilocks' region which is not too close to the star (which would mean any water evaporates away or stays as a vapour in the atmosphere) and not too far away (where any water would just remain frozen solid). At least, that was the traditional view of the habitable zone until the icy moons of the gas giants in our solar system showed that liquid water could easily exist much further away from a star than anyone had ever thought and being heated by a star is not the only way water can keep the energy to remain liquid.

Our story of the search for extra-terrestrial life, however, begins nearer to home.

A full Moon is an astronomer's worst enemy. Hanging brightly in the night sky, its reflected sunlight can easily obliterate the stars, especially contaminating the observations made by modern, ultra-sensitive ground-based telescopes. Operators of these instruments, built at great expense on tops of mountains and in remote deserts in order to avoid light pollution so that they can study the vanishing traces of celestial objects from the other side of the universe, do whatever they can to avoid looking at the Moon, for fear of a wasted evening's observations.

On the night of 22 October 2009, however, astronomers at some of the world's most advanced ground-based observatories – the Allen Telescope Array in California, the Keck and Gemini North observatories in Hawaii and the Apache Point observatory in New Mexico among others – had cast aside their usual anti-lunar sentiment and fixed their instruments firmly on the Earth's companion satellite. Some 385,000km away, a remarkable feat of co-ordinated spaceflight, two years in the planning, was approaching its final few minutes and the telescopes on the ground were ready to record every detail.

At the Moon's south pole, the Lunar Crater Observation and Sensing Satellite (LCROSS) had just split into two pieces, sending a projectile (the spent upper stage of an Atlas V rocket) hurtling at 2.5km per second towards the surface. A few minutes later, the satellite itself followed in its wake, its suite of cameras trained on the impact site to take measurements of whatever debris the projectile kicked up. Coming around the Moon at the same time, at a steady 1.8km per second, was the Lunar Reconnaissance Orbiter (LRO), timed to reach a position above the impact site within 92 seconds of the projectile hitting the surface. Tens of thousands of miles away, the Hubble Space Telescope had emerged from Earth's shadow and fixed its eyes onto the site of the collision. Millions of people back on Earth – astronomers and others – were watching too, as LCROSS hurtled through the explosion of debris and beamed live video back to Earth on its suicide mission towards the lunar surface.

The whole operation was being co-ordinated from a

control room at NASA's Ames Research Center at Moffett Field in California. LCROSS principal investigator Anthony Colaprete and mission scientists Jennifer Heldmann and Kim Ennico Smith had split the tasks of being in constant contact with almost a dozen observatories from Hawaii to South Korea and the flight operations team at Ames, who were sending instructions to the probe as it fell to the surface. 'There was an impact and we were trailing for four minutes with our instruments,' says Colaprete. 'We didn't know what we were going to see.'

In the final few moments of the descent, Colaprete was looking at the video feed from one of the infra-red cameras on the probe and wondered if they might get a better image if they turned up the exposure – at that point their screens were dark, not unexpected given that they were heading into a black crater. He asked Ennico Smith to call into the control room and ask for an exposure change on the near-infrared (IR) camera, which had a call sign 'November'. The voice from the command in the control room responded, calm and business-like, 'So that is, camera Mike 1, gain setting 3'. Somehow they had misheard or misunderstood, thinking that the science team wanted to turn up the exposure on the medium-infrared camera (call sign 'Michael'). Meanwhile, the surface of the Moon was fast approaching. As Colaprete remembers it, Ennico Smith noticed the mistake immediately and started yelling into the microphone: 'November, November, November!'

The control room turned up the correct camera's exposure just in time. The image on the screen in the Ames control room became saturated, 'then all of a sudden the middle started to peel away and started to clear and we could see the bottom of the crater with the near-IR camera, we could see it coming at us,' says Colaprete. 'That decision and interaction with the mission operation centre occurred over a thirty-second span that seemed like forever.'

LCROSS took photographs, tasted the chemicals in the dust, sent back copious data and, just four minutes into the science phase of its 100-day mission, smashed into an uncountable number of pieces on the Moon.

'As soon as it impacted, everyone was clapping and celebrating,' said Colaprete. The years of careful co-ordination, mapping and precision planning had worked – the probe had hit the Moon within 80m of its intended target. For the science team, though, the real work was just starting. NASA had scheduled a press conference for an hour after the impact so that the scientists could report on the immediate results of what had become a much-anticipated mission by members of the public and media.

LCROSS and its impactor had been smashed into the carefully chosen south pole region of the Moon to work out whether there was any water there. Previous scans of the area had indicated an increased amount of hydrogen, though it was unclear what form that hydrogen took. LCROSS was designed to answer that question and work out what lay in the soil.

At the press conference, everyone was asking Colaprete the obvious question: had he seen water? 'I remember my answer was, "I see squiggly lines."'

Colaprete was referring to the data from the probe's spectroscopes, which shone lasers through the plume of debris and looked at the reflected light in order to work out what the dust was made from. Chemical elements have characteristic signatures in these spectroscopic images because each one absorbs light at very precise frequencies. They betray their presence in the form of frequency bands missing from the light reflected back to the spectroscope.

One reporter caught on and asked Colaprete if he had seen absorption lines. Colaprete restated his earlier answer about 'squiggly lines' and gave no more away. What he didn't say at the time, but already knew, was that he had seen the preliminary spectroscopic readings from the plume of lunar dust and, in it, the characteristic light signature of water. At that moment, Colaprete was the first person in history to know for sure that the Moon had ice at its poles.

A month later, with independent evidence collated from Lunar Reconnaissance Orbiter, Hubble and the two ground-based observatories, which had all been so carefully planned to either be in or looking at exactly the right, tiny part of space on the Moon at exact, split-second timing to witness the remarkable four-minute experiment, NASA announced to the rest of the world that they had found frozen water on the Moon.

Given its proximity to home, it took a surprisingly long time to confirm that there was water on the Moon. There are no lakes, ponds or rivers on our nearest astronomical neighbour but, instead, big chunks of ice at the poles and water molecules and hydroxyl ions (made from one hydrogen and one oxygen atom) locked into the upper layers of the lunar soil. And all of that was only confirmed in the final few months of 2009.

A month before the LCROSS mission spectacularly crashed into the Moon, a separate series of experiments had reported that the top few millimetres of lunar soil contained water molecules and also hydroxyl ions bonded to the minerals there, and that there was more of it at higher latitudes. Those results had been a long time coming. The earliest hints had come from a flyby of the Moon by NASA's Cassini probe, on its way to Saturn in 1999. As we will see, ten years later, that discovery was confirmed by NASA's Moon Mineralogy Mapper (known as 'M3'), which was flying on board the Indian Space Research Organization's Chandrayaan-1 spacecraft. M3's on-board spectrometer recorded the infra-red parts of the sunlight reflecting from the Moon and split that light into its component wavelengths, in order to build a picture of what was on the surface. The analysis showed missing wavelengths of light coming off the surface that were consistent with the light absorbed by water molecules and hydroxyls.

How much water there is in total on the Moon is still uncertain but one of the M3 scientists said at the time that it could be as much as 1,000 parts per million. That would mean harvesting a tonne of lunar soil would yield just under a kilogram of water. The M3 findings were further confirmed when NASA's Epoxi probe flew past the Moon on its way to the comet Hartley 2 a year after M3 made its recordings, showing that most of the surface was hydrated, at least for part of the lunar day.

The story of water on the Moon wasn't always so clear. Though it was one of the first planetary objects Galileo studied with his telescope in 1609 and it is the only other world upon which men have walked, the presence of water there has been a cause of much confusion and debate.

The Moon formed several billion years ago from the impact between Earth and a planetismal that is thought to have been around the size of Mars. The resulting debris knocked off our nascent planet was trapped by the Earth's gravity and eventually coalesced into the Moon we see today. Not only is the Moon made of the same basic materials as the Earth, it has also shared our history, being bombarded by similar asteroids and comets, including the types that deposited water onto the Earth and gave us our oceans. The modern-day incarnations of the Moon and Earth show us, however, that the two objects had different fates even with the same starting materials. While the Earth flourished with water and life, the Moon is a dead object. This is partly because its gravitational field is only a fifth of the Earth's and it has no appreciable atmosphere. Any of the light elements raining down upon its surface that did not escape the feeble gravity would have been exposed to the vacuum of space as well as a heavy dose of solar radiation – all reasons why relatively little survived on the lunar surface.

By the mid-nineteenth century, writes Columbia University astronomer Arlin Crott, scientists had largely come to the conclusion that the Moon was dry and airless, based on the absence of any observable weather. They also realised that any significant body of water on the Moon's surface would quickly disappear into the vacuum of space.

That view was largely unchallenged until the middle of the twentieth century. Around the late 1940s, a few scientists realised that, because the Moon's axis of rotation was tilted only very slightly, by a minuscule 1.5 degrees from vertical, some of the deep craters on its surface would never see the light of day. 'People started to conjecture that there might be permanent shadows at the poles of the Moons,' says Colaprete. 'They would be very cold and, by that virtue, they could retain water even in the vacuum of space. That was conjecture, based on geometry and what we could see from telescopic observations of the mountains and craters at the poles of the Moon.'

The most significant intervention came from Nobel laureate Harold Urey in the mid-twentieth century. He won his prize for chemistry in 1934 for the discovery of deuterium (the heavier isotope of hydrogen which has a neutron as well as a proton in its nucleus). Urey wrote about the Moon as the 'Rosetta Stone of the solar system', something that might hold clues to the early years of the Earth, and was instrumental in persuading NASA to explore the Moon scientifically. Pictures sent back from NASA's Ranger 7 and Ranger 9 satellites in the 1950s and 1960s show what looked like channels that could have been cut by rivers and Urey argued that they were the results of water on the Moon's surface.

Urey came under considerable attack for his ideas, however, and he spent a lot of time defending himself from fellow scientists. In a letter to *Nature* in 1967, he said that though the subject of water on the Moon had been received mostly with great scepticism, in his view, evidence supporting his ideas had recently become quite overwhelming, particularly due to the most recent pictures sent back by Lunar Orbiter 4 and 5, probes designed to map the Moon ahead of the Apollo landings. His letter also suggested that some of his colleagues were none too complimentary about whether or not Urey came up with his ideas under the influence of alcohol. 'Because many people are not aware of this evidence and suggest that the effects are caused by other liquids, that is, lava, dust-gas or possibly even vodka, a brief discussion of the evidence may be in order.'

The first samples from the Moon came back to Earth in the 1960s and 1970s, with the Apollo missions and also robotic landers sent by the Soviet Union. The rocks seemed so dry that even Urey abandoned his ideas. He died in 1981, well before the pendulum on the evidence for the Moon's water eventually swung back in his favour.

The Apollo samples that came back, all 380kg of them over the course of the programme, generally showed very desiccated material with water concentrations that were barely measurable, somewhere in the tens of parts per million. That was consistent with the idea that the Moon was dry and if water or hydroxyl existed there it was minimal. There were some samples that showed slightly higher concentrations – up to 1,000 parts per million – but the ratios between deuterium to hydrogen in those were a match for Earth, which pointed to terrestrial contamination. Water contamination from Earth is the bane of a planetary scientist's life because there is just so much of it on our planet, everywhere. 'Even if you bring dirt and you cook and dry it and you put it in a vacuum, it still shows signs of water,' says Colaprete.

By the time the Apollo rocks had been studied, the world

seemed convinced the Moon was dry. At least, that's the American side of the story.

The Russians continued their lunar research programme well after the Apollo missions ended and, in August 1976, launched Luna 24 on a Proton rocket. In just nine days, it landed on the Mare Crisium, spent a day collecting samples from two metres under the surface and, two weeks after it had left the Earth, re-entered over Siberia. The scientists who analysed the soil found that it contained 0.1 per cent water by mass and they found more water in the samples that were deeper under the lunar surface. Their work was published in a Russian-language journal, however, and ignored by the rest of the world. Arlin Crott writes: 'The authors point out that the sample shows no tendency to absorb water from the air, but they were not willing to stake their reputations on an absolute statement that terrestrial contamination was completely avoided. Nonetheless, they claim to have taken every possible precaution and stress that this result must be followed up.' No one ever did.

Fast-forward to 1988 and one of Colaprete's colleagues, Faith Villas, was studying stony asteroids, known as chondrites, on the Moon and in particular the hydrous materials on them. 'We know that chondritic meteorites have water and have been altered by water, these OH bonds in them,' says Colaprete. 'Villas was making observations of the Moon and she was looking at observations near the poles and she saw water and hydrous signatures in the reflectance spectra. She presented this at a workshop and people brushed this aside, saying, no, you didn't do something right. She couldn't get it published for a long time.' Her work did eventually get released, but only in a relatively obscure Japanese journal in the late 1980s.

In the 1990s, there were two missions sent to the Moon to look for water. In 1994, the US Department for Defense funded Clementine. 'One of the measurements they made was a bistatic radar measurement, where they shone a radar from Earth onto the south pole of the Moon and they picked it up by the receiver on the spacecraft,' says Colaprete. 'Water ice has a very distinct radar reflectance.' Clementine saw something interesting, a south pole crater known as Shackleton, and it evoked the idea, which had first been postulated more than forty years earlier, that cold craters might hold water ice.

Four years after Clementine, a mission called Lunar Prospector was launched by NASA and managed by the Ames Research Center. It had a variety of cameras and spectrometers, one of which was a neutron spectrometer that could measure the neutrons scattering up from the Moon. As cosmic rays rain down on the Moon, they interact with the atoms in its surface soil and kick out neutrons. Those neutrons are then slowed down, absorbed and their energies are altered by the presence of certain things, especially water in the soil. This is a familiar and well-understood phenomenon: nuclear reactors on Earth are surround by tanks of water because water is so good at slowing down and absorbing neutrons.

'It showed, unequivocally, enhanced hydrogen deposits at the poles of the Moon,' says Colaprete. 'They couldn't tell if it was water ice, all they could tell was that it was hydrogen. That hydrogen could be in the form of absorbed water, OH on the clay or it could be water ice. Or it could be methane. They knew there was increased hydrogen at the poles of the Moon and it was likely associated with these permanently shadowed craters but the footprints of the instruments were so large, they couldn't be sure.'

They had their best indication yet that the Moon had the ingredients for water, and they had a region that looked like there might even be visible water molecules, if they could go and look. Now they had to work out a way to confirm it.

In the first decade of the twenty-first century, the discovery of water on the Moon became definitive. NASA's M3 instrument, on board India's Chandrayaan-1 probe, was never meant to look for water, but its surprise discovery is now what it is best known for.

M3 was a spectrometer that could operate in visible and infra-red wavelength range, from 0.4 to 3.0 microns (millionths of a metre), and could sort the light coming from the Moon's surface into any of 260 channels. As it passed over the Moon and recorded the reflected sunlight, M3 separated out the frequencies in the light and built up separate maps of the surface for each of the 260 channels. 'While the builders of M3 were aware of issues concerning water, the wavelengths covered actually cut off amid the 2.9-micron hydration band,' writes Arlin Crott. 'The amazing result from M3 was that hydration seemed ubiquitous across the lunar surface, stronger near the poles.'

The 3-micron cutoff in the instrument made working out the exact nature of the hydration difficult, though. Was it H_2O or OH? These look different to a spectrometer sensitive enough to spot the difference. The confirming measurements from Epoxi and Cassini agreed that the hydration occurred everywhere over the Moon but did not do anything to resolve the water versus hydroxyl issue.

'Those were the three papers that got published in 2009 saying, lo and behold, there's water all over the Moon,' says Colaprete. 'This is a veneer of water and we really don't know a lot about it still – there's a lot of conflicting theories about how it was formed, how it is retained.'

More recent analysis of the M3 data tells a different story. 'In some of these places, that water signature is not temperature dependent, time of day dependent; it seems associated with geomorphological edifices, like a central peak to a crater where there's a particular mineral type,' says Colaprete. A recent paper in *Nature* even suggested that this water is actually originally from the Moon's formation, inside the magma that bubbled to the surface in its early years and cooled into rock.

The M3 result was a surprise but perhaps should not have been. 'We should be clear,' writes Crott. 'Before Chandrayaan-1 and separate from radar, neutron or infra-red spectroscopy we knew that lunar water exists and came from inside the Moon. It is in the rocks. Despite scepticism about hydrated Apollo samples, we knew by 2008 that surprisingly large amounts of water/hydroxyl are locked in some lunar minerals.'

That emerging knowledge came partly from work to re-analyse Apollo samples in the early 2000s, using better equipment to delve ever-deeper into their innards. The volcanic glasses inside some of these rocks, for example, contained water. 'The isotopes tell a very complex history,' says Colaprete. 'The Moon's water cycle that we've seen with those instruments is complicated. There's water coming out of the interior, there may still be. There's water being produced at the surface because they seem to see temporal changes of the surface reflectance. The Epoxi measurements showed quite clearly that there was a day and night variation in the amount.'

The lunar day is equivalent to around twenty-nine Earth days and daylight temperatures reach 120°C. Any bulk water exposed to these conditions would quickly disappear into the vacuum that surrounds the Moon, or be bombarded with so much high-energy solar radiation that it breaks apart, with the hydrogen probably slipping the lunar gravitational bonds and escaping into space.

The poles, however, are a different matter. Recall the ideas that first emerged in the 1940s about the Moon's axis of rotation: it is almost vertical, and there are craters here that may not have seen sunlight for billions of years. Inside them, temperatures could drop to -200°C and anything that wanders into them has the potential to get trapped for a very long time, remaining there in stasis. If there were any water molecules sitting recognisably on the surface of the Moon, the poles were a good bet for a place to find them.

In 2005, NASA was building the Lunar Reconnaissance Orbiter (LRO), a mapping satellite that would survey the entire Moon in high resolution, part of the preparation for President George Bush's grand vision to send people back to the Moon and build permanent settlements there.

In the planning for LRO, NASA managers decided they needed a slightly bigger rocket than they had initially thought in order to send their payload into space. But rockets only come in set sizes and the next step up, an Atlas V Centaur, left them with 1,000kg of spare capacity. Whether or not they sent anything up they would have to pay for the extra power, so NASA announced a competition for quick-turnaround ideas to fill the extra space.

One of those ideas was the Lunar Crater Observation and Sensing Satellite (LCROSS), a relatively inexpensive mission designed and built in less than three years to test whether or not the hydrogen seen at the poles by Lunar Prospector was really water ice. 'This was important because, if it was water ice, it could be scientifically interesting and important, but incredibly valuable as a resource. It takes $10,000 to bring a pound of anything to low-Earth orbit, about three times that to bring it to the Moon and land it. When 80–90 per cent of liftoff is just fuel, if you can find a means to generate that fuel beyond the gravity well of the Earth, you can do some really great things.'

Finding water ice would be a benefit because it could be easily processed on the Moon. We know there is oxygen in the lunar soil and it is possible to make water out of it by reacting hydrogen with it and warming it up to very high temperatures – but that takes a lot of energy. 'It's easier if you have water ice in the dirt, all you have to do is warm up the dirt a little bit and you get steam,' says Colaprete. Once you have liquid water, you can do lots of things with it – drink it, use it for radiation shielding or electrolyse it to get back hydrogen and oxygen, ingredients for rocket fuel. 'If we want humans to go for extended durations beyond Earth, a resource like water could be incredibly enabling,' he says.

As we have seen, LCROSS was designed to detect water in a brute force way, by slamming itself into the southern pole of the Moon and examining what came out. In its operational phase it would be composed of two parts – the spent upper stage of the Atlas V rocket, the Centaur, to act as impactor and a Shepherding Spacecraft that would guide this impactor and carry the scientific equipment needed to take measurements of the resulting dust cloud.

The 2m-long, 600kg LCROSS left Earth (along with LRO) on 19 June 2009 and took four days to get to the Moon, where it entered an elliptical commissioning orbit for sixty days while the team of scientists and engineers back on Earth switched on and tested the spacecraft's equipment. Once that was complete, the probe was kicked into a circular orbit around the poles, 50km above the surface.

After cruising through space for 113 days and 9 million km, it reached its target location on the Moon, above a 100km-wide crater that was in permanent shadow called Cabeus. At the southern pole of the Moon, Cabeus was known to contain an area with an enhanced hydrogen signature.

On 9 October, around 4.30 a.m. local time in California, the two-tonne Centaur projectile was sent crashing into Cabeus. The Shepherding Spacecraft followed four minutes after the Centaur and recorded the impact and debris plume with three spectrometers, five cameras and a photometer looking down at the surface of the Moon as well as up at the Sun to examine the debris that was thrown up. At the same time, LRO observed the impact, as did the Hubble Space Telescope and a variety of other ground-based telescopes in Hawaii and the western United States. 'We covered it in every possible way,' says Colaprete. 'We wanted to go in with our eyes wide open. We didn't want to presume it was water and have all of our assets focused on just seeing water and then miss it because it was something else.'

NASA had crashed rockets into the Moon before. During the Apollo programme, the Saturn V upper stages were ditched into the Moon as part of their normal trajectory. Later missions used these crashes for scientific purposes. 'They had seismometers installed from previous missions and they would deliberately slam the upper stage into a particular place on the Moon to watch the seismic waves,' says Colaprete. 'There was a history of doing this but not in the way we were doing it.'

When his team had contacted the ground-based observatories to ask for assistance, the astronomers there had told him, in no uncertain terms, that looking at the Moon was a tough ask. 'When we asked them to make these observations, they said, "We don't even know how to point at the Moon. We avoid it,"' says Colaprete.

And, to make things worse, how do you accurately point the ground-based spectrometers to pinpoint the tiny spot on the Moon where the impact was likely to occur? Telescopes usually use the stars to orient themselves but that was not an option here. Their solution was to work with a lunar geologist who could differentiate the Moon's craters. 'To the untrained eye, one crater looks like another crater [but] they developed a methodology where they hopped from crater to crater like they were guide stars to get their spectrometer exactly where it needed to be,' says Colaprete.

The whole team rehearsed their experiment to make sure they were drilled for the four minutes they had to capture their data. Was he sure that this rapid feat of organised, co-ordinated astronavigation and engineering would work on the day? 'No, no, no,' says Colaprete.

It did work. LCROSS made a 30m-wide, 4m-deep crater in Cabeus, excavating 6,000kg of lunar soil into a plume that rose nearly 10km high, exposing the material there to sunlight for the first time in several billion years. Its measurements unambiguously showed there was granular, crystalline water ice in the shadowed crater, at a concentration of around 5.6 per cent (+/- 2.9 per cent) by mass.

'We also saw a whole load of other volatiles come up; we saw light metals like sodium, mercury, potassium, calcium,' said Colaprete. 'We saw gold and silver but I think that was contamination from the rocket engine. We also saw a variety of other volatiles – methane, sulphur dioxide, carbon dioxide. And we saw a lot of hydrocarbons too, such as ethylene. These cold traps were really just trapping everything, not just water ice. By far the greatest concentration of volatiles in it was water ice.'

They found water molecules in ice as well as hydroxyl ions, a product of the breakup of water molecules that had been thrown up into the plume and split apart by sunlight. Later analysis showed that the vapour cloud after the Centaur impact had been heated to 1,000 Kelvins and contained around 570kg of carbon monoxide, 140kg of molecular hydrogen, 160kg of calcium, 120kg of mercury and 40kg of magnesium.

LCROSS showed that there was a lot of interesting material, in particular water, at the Moon's poles. It's not exactly pleasant stuff – ice containing a heady (and probably poisonous) cocktail of metallic and organic compounds – but the principle is there, and there are ways to purify any broth that comes out if we decide to start mining it for future Moon bases.

This compelling sniff of H_2O does not complete our basic understanding of the how and why of water on the

Moon, however, and it is worth exercising some caution before getting excited about using it as a resource. 'Having denied the existence of lunar water for so long, there is a tendency to make the sudden transition to presuming that we now understand the basic story and can narrow our vision to confirming our predictions,' writes Crott. 'Not so fast.'

Many of the remaining questions will be tackled by the next generation of lunar missions. The most prominent is NASA's Resource Prospector, a 300kg rover that will take a scientific laboratory onto the Moon. It will have a neutron spectrometer very similar to that on Lunar Prospector, which made the first maps of hydrogen at the poles, but it will be out in front, effectively the rover's nose. It will land in the polar region where scientists think there is hydrogen. It will also have a near-infrared spectrometer that will look out in front at hydroxyl and water in the lunar surface. 'If we find a place of interest, the rover has a drill and it can go to a metre and take samples,' says Colaprete. 'We can do two things with that sample – we can bring it up and brush it into a pile for the near-IR spectrometer to look at and it can take a quick assay or look for ices with a light; also we have an oven that we can deposit samples into and that can warm as high as 900°C. Typically we'll warm it to 150°C and hold it there, and any vapours that come off the sample will go into a tank that we will then look at with a transmission near-IR spectrometer. We can do a simple transmission measurement for water and other gases.'

This suite of instruments on the rover will go into both lit and shadowed areas of the Moon and give us the first hands-on ability to study the surface, in the way a human geologist might work. The results will be crucial for any plans for future human returns to the Moon, which might be long-term and therefore require the mining of materials and resources on the surface.

We can learn much from our Moon. There has been no weathering to destroy rocks there, unlike the Earth, and it is therefore a time capsule for the early solar system. Any astronomical events that have happened to the Earth since its formation have happened to the Moon, and the Moon has preserved that record better. Water is a big part of that story because planetary scientists can look at the isotopic fractions of any ice – how much hydrogen there is in the water as opposed to the heavier deuterium – and this can tell them not only what has been hitting the Moon in its history, but also where it came from.

We know that the moon has water but a big open question is how that water got there. One way H_2O might have appeared in the soil is when the copious hydrogen that streams out from the Sun hit the Moon and got trapped. Or the water could have been delivered by comets or watery asteroids. Or it was already present in the rocks gouged out of the Earth. Probably all of these scenarios are true. In that last regard, water is a window into how the Moon formed and cooled, how the molten magma ocean at its surface in its early years spread and solidified into what we see today.

The distribution of any water (as ice or bonded into the soil) gives an insight into the history of the Moon. 'The Moon is interesting because its distribution is complex – not all cold craters show signs of hydrogen or water. Cabeus, where we impacted, showed one of the highest concentrations on the Moon. It is cold: where we impacted the temperatures were 40K [-233°C]. Other craters that were almost as cold, say 70K [-203°C], don't show any of this water or enhanced levels of hydrogen. Some show potential frosts at the surface.'

Why the water has this distribution is a mystery. One idea is that the Moon has, through its history, exposed different craters to different amounts of sunlight as its orbit around the Earth has evolved. 'Some modelling that's been done shows that it affects some craters more than others – it just so happens that Cabeus is one that was in a state favourable to retaining water on the surface most recently. What we could be looking at on the Moon is a potentially massive comet impact in the last billion years and it went to all craters, but only certain craters retain it to today where we can see it,' says Colaprete.

The prospect of studying the ice at the poles is a potential scientific goldmine, a bit like taking ice cores on Earth. If scientists could get to the Moon, to one of the places where there is potentially thick, blocky ice in the subsurface, they could take a core of that and be looking at a comet that is a billion years old. Within that would be a history of the solar system.

'We've really moved into the next stage of the lunar water story,' says Colaprete. 'We know there's water ice and we know more or less where it's distributed. But there's no one theory yet that explains why it's distributed the way it is, why it's at the concentrations it is. We don't know how deep it goes.' He means that physically and metaphorically.

And the Moon, of course, is only the first waypoint in our search for water – and ultimately the signs of life – further afield in the solar system.

TEN

Mars

On 5 August 2012, just after 10 p.m. local time in Pasadena, California, a room filled with engineers and scientists at NASA's Jet Propulsion Laboratory (JPL) fell silent. Mars Science Laboratory (MSL) had just separated from its cruising rockets and entered the red planet's atmosphere after a nine-month, 300-million-mile journey from Earth. Over the next seven minutes, the spacecraft would use pre-programmed instructions to guide itself through a complex set of manoeuvres to fall through the planet's thin atmosphere and deliver its payload, the Curiosity rover, to a precise spot on the planet's surface.

Years of meticulous research, design and testing by NASA engineers had gone into planning that landing sequence. Thousands of planetary scientists hoped to use Curiosity to work out whether or not Mars was a planet fit for life and the survival of their delicate experiments rested on what was going to happen in the following seven minutes. There was nothing that the team back on Earth could do to intervene at this point, nothing but wait. MSL was alone. No wonder NASA dubbed the landing sequence 'the seven minutes of terror'.

Even by the standards of our modern era of space exploration, where we've become accustomed to a regular diet of spectacular pictures and video from other worlds and unprecedented precision in our ability to examine our solar system, Curiosity's arrival onto the red planet was a gobsmacking, audacious piece of engineering.

As the MSL's descent stage smashed into the top of the Martian atmosphere at 21,000kph, explosive charges jettisoned two blocks of tungsten, each one weighing 75kg, to shift the orientation of the spacecraft so that it could fly controllably through the Martian air. Aerodynamic drag quickly pushed the outside of MSL's heat shield to a temperature of 1,600°C while its on-board computer used small thrusters to guide the spacecraft into a series of zigzags in the air, not only to slow it down but also to line it up precisely for its landing spot on the ground at a place called Gale Crater. As it fell, MSL ejected six more 25kg tungsten blocks to help keep itself pointing in the right direction. When its aerodynamic manoeuvring had slowed it to 1,600kph, the falling spacecraft deployed the biggest and strongest supersonic parachute ever built by NASA, which, despite weighing only 45kg, could withstand up to 29,000kg of force. When that opened, the sudden deceleration jerked MSL by nine times the force of gravity on Earth.

Next, explosive bolts jettisoned MSL's heat shield and exposed the underside of the lander, where a camera began recording the landing at four pictures per second and an on-board radar began to take altitude and velocity measurements that the computer could use to make any necessary adjustments for the final moments of the landing sequence.

The parachute reduced the speed of the falling MSL to around 320kph by the time it was cut off, 1.5km above the surface, but that was still not slow enough for the spacecraft to land without being smashed apart. To slow down further, MSL fired eight retrorockets at the Martian surface until it was just inching towards the surface. As it did so, it carried out a sharp turn to move itself out of the way of the falling parachute.

At around 20m above the surface, the spacecraft hovered and opened its hold doors to expose the 3m-long, one-tonne Curiosity rover. The size of a small car, this was the largest NASA lander yet. The space agency's tried and trusted methods to cushion the impact of a landing from space – air bags, parachutes and retrothrusters – were not strong enough to land Curiosity all by themselves. Engineers had come up instead with another daring manoeuvre to get Curiosity to the ground, which they called the 'sky crane'. As the spacecraft hovered above the surface, Curiosity was lowered the rest of the way to the surface on three 6m-long nylon cables. (Even Adam Steltzner, in charge of the space-craft's entry, descent and landing team that designed the sky crane, admitted that the manoeuvre looked 'a little crazy'.)

Once that was complete, the rest of the spacecraft's descent stage executed its final pre-programmed command, firing its rockets one more time to move away from the rover and crash-land a safe distance away.

It takes around fourteen minutes for a signal to travel, at light speed, from Mars to the Earth. As soon as the team at JPL had received the message from MSL that the space-craft had entered the top of the Martian atmosphere, they already knew that, at that exact moment on Mars, Curiosity had already been on the ground for seven minutes. Whether the rover was functioning or whether it was in a million pieces, strewn across the floor of its target landing site at Gale Crater, they had no idea. In the JPL control room, the seven minutes of terror manifested itself as a total, deep silence. Via websites and live television feeds, the rest of the world held their breath.

At 10.32 p.m. local time in California, a signal arrived

at the consoles in the JPL control room, confirming that Curiosity had landed safely on Mars. The scientists and flight engineers in Pasadena erupted into hugs, cheers and sobs.

All of the inner planets of the solar system have (or have had) water in or on them. Formed, like the Earth, from the coming together of ice-encrusted dust grains and boulders 4 billion years ago and then bombarded with water-containing comets and asteroids during the Late Heavy Bombardment, all of them started life with water. The subsequent history of each planet and its water, however, has proved wildly different.

Closest to the Sun and bombarded by intense heat and UV radiation, Mercury never had a chance of hanging on to any of its liquid water. The smallest planet in the solar system, its surface is baked to more than 400°C and it has never seen seas or an atmosphere – we know that because either of those things would have smoothed the heavily cratered ground over time.

It seems to have hung on to some of its ice, however. When NASA's Messenger probe reached the planet in 2012, it found evidence for water ice and also organic chemicals. Both chemicals were found at the poles, at the bottom of deep craters that are in permanent shadow. The ice survives because, as Mercury circles the Sun, its axis of rotation is tilted less than a degree from being perfectly upright, which means the planet never points its poles towards the Sun and the craters there remain dark and cold. If the planet has been orbiting like this ever since it was formed, that ice could be truly primordial, frozen since before the birth of the solar system as we know it.

Venus, a virtual twin of Earth in many ways, had plenty of water when it was formed. But, as the Sun got older and became hotter, it suffered a runaway greenhouse effect because of the large amounts of carbon dioxide in its atmosphere. This created a hothouse planet where, between its formation and now, any oceans boiled up and eventually leaked away from the planet.

After the Moon, Mars has the most well-understood story of water in the inner solar system. We've been sending landers, orbiters and examining meteorites from this planet for decades, trying to do anything we can to work out whether there was liquid water on its surface, what happened to it and, by extension, where we might find hints that there could once have been life. The picture we have built up is, at the same time, frustratingly incomplete and yet the most detailed we have of any world other than the Earth. Here's what we know.

In its early history, Mars had liquid water just like the Earth and Venus. 'We think it was Earth-like in its environment, with water on the surface flowing. The compelling thing about early Mars is that, as far as we can tell, all of the environments that would have been present on the early Earth would have been present on the early Mars,' says Chris McKay, an astrobiologist at NASA's Ames Research Center in Mountain View, California. Around 3.5 billion years ago, the baby versions of both our planet and Mars were near twins. Both had liquid water on their surfaces, plenty of sunlight, warmth and other clement conditions for life to begin.

Microbial life on Earth started quickly once the tumult of the Hadean era had settled down and subsequently evolved, over billions of years, into the myriad complex forms we see today. But Mars seems to have gone in a different direction. 'Not only does it die but it appears that any record of life, at least on the surface, has been oxidised and irradiated to smithereens,' says McKay. What scientists want to know is, given the window of similarity between Earth and Mars early on, around 3.5–3.8 billion years ago, was there a chance that the red planet also had simple life for a while in certain regions? Might they exist, in rare amounts, today?

And even if life never existed there, is Mars in any way habitable enough for humans to colonise it one day?

Ever since Galileo spotted Mars through his telescope four centuries ago, people have probably wondered what the environment of this red planet is like, whether there was water, whether there was life there. The first time we saw images of Mars taken from a spacecraft was in 1965, when Mariner 4 made a flyby of the planet. The scientific community had abandoned the idea of finding intelligent life living on the planet by then but were unsure about whether microbes or forms of vegetation might still exist somewhere on the surface. Outside that select group, meanwhile, people were excited and open to possibilities, given that the images of the surface thus far had showed what looked like canals and valleys.

The dozen or so Mariner 4 pictures beamed back to Earth stopped all of those extravagant dreams of life in their tracks. What the world saw, even in those grainy images, looked more like the Moon than the scene of anything vibrant and living: big impact craters, volcanoes and arid-looking plains. Certainly no signs of canals, vegetation or anything like that.

In 1972, Mariner 9 became the first spacecraft to get captured into Mars's orbit, rather than just flying past. It arrived just as a dust storm was raging, blocking any view of the surface. When that cleared, Mariner 9 sent back higher-quality pictures of the craters, valleys and high plateaux it saw underneath.

'That's when all these features that signify ancient water were discovered,' says Ashwin Vasavada, a planetary scientist at NASA's Jet Propulsion Laboratory in Pasadena, California, and a senior scientist on the MSL mission. There were two main sets of features that got everyone excited: huge channels that looked as though they had been created during catastrophic floods at some point in the planet's history and networks of valleys that were smaller but more numerous. The valleys, evidence of ancient rivers, pre-dated the outflow channels and appeared on some of the oldest regions on Mars. The southern highlands, for example, had hundreds of valley networks flowing from high to low ground, some of them a thousand kilometres long or more. But these were all remnants of water, whereas everything Mariner saw was dry and looked like it had been that way for billions of years.

Just after it landed in Gale Crater in 2012, the NASA MSL mission's rover, Curiosity, discovered it was rolling around on a dry river bed, formed by an ancient river coming down from the crater wall and spreading out rocks and sand into an 'alluvial fan' at the base. On the day of landing, Curiosity found itself among rounded pebbles and other sediments that were the tell-tale signs of flowing water. It was not a complete surprise – the alluvial fan had been seen from space and was part of the reason the NASA team had chosen the site as the place to put the rover down. Mariner 9 had also seen similar alluvial fans in 1972.

The importance of this observation, scientifically, was that Curiosity was sitting and staring at the ancient river bed from just metres away and direct measurements and close-up images – ground truth – is a necessary confirmation for the images taken from above. Similar features to alluvial fans can sometimes be caused without water but the detailed images from Curiosity told the scientists back on Earth that the ground underneath their rover had been shaped by water, as opposed to a dry avalanche or a wind. Only water would have had the required momentum to carry the rocks into position and the rocks' size, shape and distribution gave clues to the speed and direction of the ancient river. It's also worth noting that this is only the third place we know of in the solar system that has rounded rocks, after the Earth and one of Saturn's moons, Titan. All in all, tantalising clues to help fill in the water history of our closest planetary neighbour.

We continue to examine Mars like never before. There are three spacecraft – Mars Odyssey, Mars Express and Mars Reconnaissance Orbiter – in orbit around the planet and two active rovers, NASA's Opportunity and Curiosity, trundling around on its surface. (Opportunity's twin, Spirit, lost contact with Earth in 2010 after getting stuck in sand.)

We know that water exists on Mars, in the form of ice at the poles and vast glaciers at the bottoms of shadowed craters, but our knowledge of its full extent below the surface is still elusive. And nothing permanently flows on the surface, as far as we know. Though evidence from Mars Reconnaissance Orbiter published in 2015 suggests that there could be seasonal flows of water running down particular sloping parts of the surface, when they are warmed by the Sun in the Martian summer. Because of low temperatures and atmospheric pressure on Mars, any liquid on the surface would soon freeze and sublimate (turn directly from solid to vapour when a little heat is applied) into the atmosphere and, eventually, float out into space.

Over the past forty years, the combined American, Russian and European missions to the planet have gathered increasing evidence for a complex water history on the red planet.

Mars Global Surveyor (MGS) undertook comprehensive mapping of the planet's surface in the final few years of the twentieth century and showed that Mars was a geologically varied place with different amounts of water and weather in different places, resulting in all sorts of different potential habitats, one of which might have been suitable for life at some stage in history.

The curse of Mars struck with NASA's next two missions to the planet – Mars Climate Orbiter and Mars Polar Lander

– neither of which made it to the planet intact due to a combination of bad luck and human error. The curse is ever present with scientists who want to study this planet and a reminder of just how complicated it is to send something safely to a location so far away – more than half of the missions attempted to the planet in the past four decades have failed in some way. The UK's lander, Beagle 2, landed on the surface of Mars on Christmas Day 2003 and even deployed some of its solar panels, but it never got back in contact with Earth. The Russian Phobos-Grunt never made it out of Earth's orbit in 2011.

NASA had better luck with Mars Odyssey. Measurements from that satellite of a large amount of hydrogen buried around the poles, together with data from MGS which showed that the southern polar ice cap could not be pure carbon dioxide as previously thought, pointed to a large amount of water ice in the planet's poles covered in a layer of frozen carbon dioxide, though some of the water also looked as though it might be exposed.

That was followed by a highly detailed geological map of the Martian surface produced by Mars Reconnaisance Orbiter's high-definition cameras and spectrometer. These maps were so detailed that planetary scientists now know more about the red planet's geology than geologists do about the Earth's, which is largely covered by vegetation and oceans. Martian geology, which is mostly exposed when the wind isn't whipping up storms in the atmosphere, was clear to see for MRO to record and its maps became crucial for selecting sites on the ground for further study.

The Mars Exploration Rover mission that deposited the Spirit and Opportunity vehicles onto the surface is based on the data from the Odyssey and MRO probes, among others. These two workhorses arrived on the surface in 2004 and immediately began sending back jaw-dropping colour images of the Martian vista and valuable measurements of the soil itself.

Opportunity, in particular, pushed the water story forward when it found a mineral near its landing site in Meridiani Planum known as grey haematite, whose formation is linked to the presence of liquid water. Around the same time, the European Space Agency's Mars Express orbiter confirmed that the planet's southern ice cap – with a volume of up to 3 million cubic kilometres – was mostly water ice with a thin layer of solid carbon dioxide on top (the northern ice cap was confirmed to be water ice in the late 1970s by NASA's Viking landers). That ice at the south pole is probably just the start – the measurements pointed to a vast amount of ice buried under the surface at the southern pole, in a region known as the polar layered deposits.

'There's a tonne of water, by some people's calculations it would cover the whole planet by 20 or 30m if you melted the polar caps and unfroze the permafrost as well,' says Vasavada. 'The water hasn't been lost, it's just no longer liquid and flowing in a hydrologic cycle.'

The story of water on Mars that has emerged from these probes, landers and rovers shows that, for its first billion years, rivers flowed on the surface and created the many valley networks we see, a time scientists call the Noahchian period (after Noah, he of the floods and the ark). By the end of that period, the rivers were gone and Mars entered its middle geological era, when various catastrophic events meant that the large outflow channels began to form and dot the surface. How these channels form is still unknown but they could be a kind of groundwater eruption, something that caused underground water to break through the surface and gush out to create these huge 1,000km-long scours in the land that probes can see from space. And then, in the past few billion years, Mars has just been dry and dusty. Except, scientists think, for a few seasonal streams in areas warmed by the Sun.

Unlocking the mystery of how and why the climate on Mars changed to turn it from a planet that could support liquid water, flowing in rivers and falling from the skies, and into the place we see today, is the task of the next generation of missions. The geological evidence tells us that Mars once had the ability to support liquid water, whereas now it would boil away or freeze.

One of the ideas for the change in circumstance starts from the premise that Mars must have had a thicker atmosphere in the past, since that (and the associated higher atmospheric pressure) would have kept liquid water stable on the surface. Between then and now it has lost a large fraction of the atmosphere, taking with it the planet's liquid water.

Mars has been losing its atmosphere to space (as, indeed, have the Earth and all the other planets we know of) since it was forged out of the planetismals in the dusty disk around our young Sun. The main reason for this is high-energy particles from the Sun, which bombard the upper layers of the air and, one molecule at a time, strip the atmosphere away. A high-energy particle hits a gas molecule in the atmosphere, excites it and it gets the energy to escape the planet's gravity and head off into space. 'This is happening today and, if you integrate backwards in time, you can account for a lot of the loss of Martian atmosphere,' says Vasavada.

If Mars has lost a significant portion of its atmosphere in its 4.5-billion-year history, why hasn't the Earth also suffered in the same way? Fortunately for the story of water (and life) here, our planet's magnetic field protects us. A significant portion of the high-energy particles from the Sun are electrically charged and our magnetic field guides these particles to travel around the Earth's atmosphere rather than into it. We see the energy of these particles dissipated in the glorious explosions of colour at the aurora borealis and aurora australis, so their atmosphere-stripping effect is somewhat curtailed.

Mars doesn't have an equivalent planetary magnetic field today and therefore suffers. However, it probably did have one early in its history. Mars is made of very similar stuff to the Earth, so why does one have a magnetic field while the other does not? The answer lies in their different sizes – Mars is about half of the diameter of the Earth and is about a tenth of the mass. That made a big difference in how fast Mars cooled from its early days, compared to the Earth, and that, in turn, had effects on the planet's magnetic properties. 'Smaller objects have more surface area relative to their mass and just the simple fact that Mars is a smaller planet means it lost more of a relative amount of its internal heat than Earth did faster and therefore its crust thickened and its mantle cooled,' says Vasavada. In order to keep a magnetic field going, a planet needs magma, metallic rocks, to be circulating deep inside it.

There is evidence of Mars's early magnetic field in little patches of magnetism left on the surface, caused by the initial churning of molten rock in its innards. As the planet cooled, the rocks stopped flowing so much and, some time in its first billion years, the magnetic field disappeared. (The Earth, by the way, is following the same path and is just less far along in its cooling. We will end up, like Mars, without a magnetic field someday.)

All of this means that the atmosphere-stripping high-energy particles have had a longer period of time to do their work. That stripping away has been going on in full force for probably 3 billion years and has accounted for a good fraction of Mars's early atmosphere being lost. Estimates suggest the Martian atmosphere was up to ten times thicker, at its peak, than it is today.

Because of the atmospheric pressure on Earth, liquid water has a big range of temperatures in which it can exist. On Mars, its atmospheric pressure is so low that there are only a few degrees between boiling and freezing. A liquid water river cannot last very long in those circumstances before it freezes or boils away. 'The bottom line is, today, because of the thin atmosphere and the cold planet, it's not a place where liquid water has a chance,' says Vasavada. 'So most of it is frozen in the polar caps.'

With almost no liquid water present, we now know that the chances of finding life on Mars that exists today are slim. But this has not stopped people from trying.

'To generations of scientists, writers and the public at large, Mars seemed the world most likely to sustain extraterrestrial life,' wrote the astronomer Carl Sagan. 'But flybys past and orbiters around Mars have found no excess of molecular oxygen, no substances – whatever their nature – enigmatically and profoundly departing from thermodynamic equilibrium, no unexpected surface pigments and no modulated radio emissions.'

In 1976, NASA's two Viking landers went to Mars, landing 5,000km away from each other on the surface and both equipped to sniff out hints of life if it had ever existed on the planet. 'One experiment measured the gases exchanged between Martian surface samples and the local atmosphere in the presence of organic nutrients carried from the earth,' wrote Sagan. 'A second experiment brought a wide variety of organic foodstuffs marked by a radioactive tracer, to see if there were life forms in the Martian soil that ate the food and oxidised it, giving out radioactive carbon dioxide. A third experiment exposed the Martian soil to radioactive carbon dioxide and carbon monoxide to determine if any of it was taken up by microbes.'

Sagan, who was a member of the research team for the Viking landers, tells the story that, to the initial astonishment of all the scientists involved, each of the three Viking experiments gave what at first seemed to be positive results.

'Gases were exchanged; organic matter was oxidised; carbon dioxide was incorporated into the soil.'

The results did not stand up to scrutiny, however, and the idea that there were life forms present could not be confirmed in subsequent searches for organic molecules in the Martian soil.

None of this precludes the idea that life of some kind might have existed at some stage in the planet's history and remnants of any life from the Noahchian period might still be detectable in the soil. Once the character and general location of the Martian water had been established by the 1990s and early 2000s – the Opportunity rover, for example, confirmed that liquid water had flowed across the surface of the planet in a sustained way in its ancient history – the focus of Mars exploration began to shift to more detailed questions about the planet's water, specifically whether it was (or ever had been) in the right conditions to support life. Never mind if Mars is inhabited now, is it or was it ever habitable at all?

MRO had found sedimentary rock formations, which had formed in the presence of water, and Curiosity was sent to one of the promising locations it had identified – Gale Crater – to examine the place further. 'With Curiosity, we asked ten other questions about water – even though water is a prerequisite for life, not all water makes a habitable environment,' says Vasavada. 'The water can be too acidic, it can have too much stuff dissolved in it, and even if you had water it may have been too short-lived.'

For habitability, an environment suitable for an Earth-type life on Mars, scientists needed to look not only for liquid water that persisted in one place for a decent amount of time so that there was a chance for life to get a hold, but the chemical ingredients that life could use (including carbon, nitrogen, oxygen, sulphur etc.) and a ready source of energy.

We know that, on Earth, life occurs everywhere

wherever there is water, whether it's at the top of a mountain, around a nuclear reactor or in a bottle of acid. Extending that logic, any place where water might have been on Mars could be a potential habitable region and, therefore, hold remnants of life. But extremes are not what Curiosity was sent to look for – the NASA team wanted to maximise their chances of finding something by choosing places that had the most clement and benign possible conditions in terms of temperature, water and protection from high-energy radiation.

'We spent years picking the right landing site where we thought all the right factors would come together to make this place have the potential to be habitable,' says Vasavada. Curiosity's route involved circling and climbing Mount Sharp, which lies at the centre of Gale Crater, a 5km-high mountain of layered rock that will help scientists unravel the history of the planet. The rocks at the top of the mountain are younger than those at the bottom so, by climbing up the slopes, Curiosity will be able to travel through geological time. It is a simple idea. Gathering that data on the ground, though, is not straightforward.

The Mars Yard at NASA's Jet Propulsion Laboratory (JPL) lies just a few minutes' drive up the hill from the main entrance. It overlooks the headquarters buildings at JPL and also the gardens near its entrance. Further down the hill it is possible to make out Pasadena and, much further in the distance, the Hollywood Hills and the permanent haze hanging over the sprawl of Los Angeles. More than 5,000 NASA employees engineer and test flight systems here, and design scientific experiments that can take measurements of stars and planets in the depths of space and reliably report them back to Earth. Astronomy, geology and engineering are the stock in trade of the space agency but plenty of the scientists here work on fundamental biology too, trying to understand the basic chemical processes of life and how it might have started on Earth – a way of helping future missions look for living things outside our planet.

About the size of a football pitch, Mars Yard sits in messy contrast to the well-kept JPL roads and buildings nearby. The yard feels like a piece of a small, dusty frontier town from an old Western which has somehow been plonked into the middle of a well-to-do university campus. The ground in the Yard, fenced off from the road on all sides, is covered in dry soil, crushed rocks and sand.

Despite this rudimentary appearance, the Mars Yard has been carefully designed as a simulacrum of the Martian surface around Curiosity. Here they test how the rover might fare in the presence of different materials under its wheels; they test parts to destruction and give advice to the teams in the control room on how to drive the rover across the oncoming landscape without damaging the vehicle.

Different areas of the yard have obstacles built into them. Piles of rocks, everything from fist-sized stones to mini boulders, sit around all over the yard. Sand traps and inclines at one end imitate soil and sand dunes. Hard, flat slate stones fixed to a slope at various angles simulate the bedrock at the foot of Mount Sharp. The engineers here cover them in sand and see how easy it is for replica Curiosity wheels to drive up them without slipping.

A large shed spans one side of the yard, near the entrance. Inside, surrounded by racks of computers and boxes of parts, is a full-scale replica of Curiosity. Maggie (MSL Automated Giant Gizmo for Integrated Engineering) is a duplicate to Curiosity in almost every way – she is complete with the same control systems, motors, robotic arm, suspension, cameras and most of the scientific equipment. This Earthbound testbed is used by NASA engineers to check not only how Curiosity drives across the Martian landscape, but also how its software and systems respond to commands from mission control. On its machined aluminium wheels it is possible to see the tread pattern, shared by its counterpart on Mars, that has become an in-joke for space fans – the letters 'JPL' in Morse code. As it travels across the surface, Curiosity imprints these letters into the soil so that the control team on Earth can look to see how far it has been and whether or not any of the wheels are slipping across the surfaces they encounter. (It also, of course, leaves behind a 'JPL was here' tag, possibly the most exclusive bit of graffiti in the solar system.)

The Mars Yard is used to solve problems in real time for the rover on Mars. 'Earlier in 2014, we started noting that our metal wheels were getting more tears and punctures than we had planned,' says NASA's Ashwin Vasavada, deputy project scientist of the MSL mission. 'We knew we would have some and they would build up over the mission but [we were hoping] they wouldn't get to the point where the wheels weren't useable before we didn't need the wheels any more. But earlier this year that rate was getting alarming; it was a little more than we had budgeted for the whole mission. We stopped and did a lot of work here and with our geologists to figure out what was causing the damage and why that terrain was more dangerous than we had anticipated.'

What they found was that small, pointed rocks on the Martian surface, sculpted by the wind, were sitting on top of the hard bedrock over which the rover was travelling. The combination was damaging the wheels at a faster rate than expected, with the result that the Curiosity team at JPL decided to reroute the rover away from bedrock and onto softer sand on its route to Mount Sharp. Just to be on the safe side, the rover photographs all its wheels every few weeks and sends back the images for analysis at JPL.

In the middle of the yard, Vasavada's team attached

three replica wheels from the Curiosity rover to a rig that drove them in endless circles on a crushed gravel track, bumping and pinging over stones along the way, just to see how long they would take to fall apart.

Curiosity's wheels are each made from a single cube of aluminium that is machined into shape using lasers. This process of manufacture makes it extremely strong because there are no seams or other obvious points of weakness where it could fail. But as the wheels interact with the Martian soil, they will eventually wear out.

On the test rig, engineers simulated the torques on each wheel and also the stresses induced by the weight of the rover on Mars (gravity there is less than 40 per cent of that on Earth). They also put rocks of different sizes in the path of the wheels so that they would have to climb over them. 'The goal of this test is to put the wheels through their entire life span. We think what's happening on Mars is that we're starting to see the beginning of getting the wheels damaged. Of course, we want to know what that means for how much time we have left. We don't want to use the one on Mars to figure that out.'

In just a week, each wheel travelled 40km on the test rig (compared to about 10km for Curiosity in its first two years). 'One day it'll break and then we'll know what to look for, the steps to failure, on Mars,' says Vasavada.

The engineers at Mars Yard can test out many potential scenarios for Curiosity, so that it can avoid trouble, but the rover itself also has the brains to avoid difficult scenarios. Avoiding bedrock and going for softer sand might help maintain its wheels but that strategy has other risks, as seen when one of the other MER rovers, Spirit, was lost when it became embedded in sand. But Vasavada says that is less likely to happen to Curiosity. 'It feels whether its wheels are getting traction or not; it'll stop the moment it feels like it's digging in. You won't get a situation that happened earlier with

Spirit, which was spinning for an hour and getting deeper and deeper. Curiosity won't do that.'

The Curiosity rover is the most capable scientific laboratory ever to land on Mars or, for that matter, any other planet. Powered by lithium-ion batteries and the radiation from a chunk of plutonium, it can sample the geology in multiple ways. A laser can burn holes in rocks nine metres away and use a spectrometer to analyse the chemicals released in the vapour. A robotic arm, capable of holding 30kg of instruments and equipment, can drill holes and smash rocks apart to get at the stuff inside. A high-temperature oven can bake samples to almost 1,000°C, sniff the gases come off and work out what the soils and rocks are made from. All of this is in the service of one overarching goal: to find out if Mars is, or has ever been, habitable.

Curiosity's landing site, Gale Crater, dates to 3.5 billion years old, around the time of the end of the Noahchian period. The surface of Mars is usually dated by counting how many impact craters there are in the vicinity. Assuming there is a constant flux of asteroids and comets hitting the planets, that means the more scars there are, the longer that bit of ground has been exposed.

What drew NASA to Gale Crater, besides the ancient river system they landed on and were able to scrutinise, was their observation from orbit that it contained clay minerals. Clays (the generic term for a whole class of minerals) are important to this story because they are made when water interacts with the primitive rocks from which planets are made. Over time, the water breaks these rocks down and clay is a by-product. 'You go to the fresh volcanoes in Hawaii and it's pristine volcanic rock; you go to the older volcanic islands on Hawaii and you've got hundreds of feet of clay soil that's built up over time,' says Vasavada. 'It's a similar effect on Mars – you see clay minerals, that tells you that water is interacting with the rocks. Those clay minerals wouldn't form if the water was too acidic so it tells us the pH of the water. By seeing clays we know it's about a neutral pH. That's also something that's in favour of life.'

A year into its mission, Curiosity confirmed that it had found clays in a sample that it had drilled from a rock, named John Klein, sitting nearby on the surface of Yellowknife Bay in Gale Crater. Again, NASA knew there was likely to be clay at the location but Curiosity provided hard evidence for the first time. The experiment had involved drilling the rock and placing the pile of grey dust into the CheMin instrument on board the rover, which fired X-rays through it and found that at least 20 per cent of it was in the form of clay minerals such as smectite. The instrument also found halite (a salt mineral that can be tolerated by life) and other minerals including feldspar, pyroxene, olivine and magnetite. Another instrument on board Curiosity, the Sample Analysis at Mars (Sam), sieved a small amount of the dust and warmed it up to 835°C in its oven. It found that water was released, consistent with the presence of clays.

The discovery of water bonded into the soils was also noteworthy, not only to reconstruct the planet's history and work out its potential for harbouring life but also as a resource for future human visits.

'We tend to think of Mars as this dry place – to find water fairly easily out of the soil at the surface was exciting to me,' said Laurie Leshin, dean of science at Rensselaer Polytechnic Institute, when the initial results from Curiosity were published in a series of papers in the journal *Science* in January 2013. She was lead author on the experiments that discovered the water in the soil. 'If you took about a cubic foot of the dirt and heated it up, you'd get a couple of pints of water out of that – a couple of water bottles' worth that you would take to the gym.' Her analysis showed that 2 per cent of the soil, by weight, was water.

It won't all be plain sailing for potential human missions, though. Leshin's analysis also showed that there was a chemical called perchlorate, which is toxic to people because it can impede thyroid function. Another early result was the analysis of a rock near Curiosity, named Jake_M after a NASA engineer, which turned out to be similar to a type of rock on Earth called a mugearite, something that is typically found on ocean islands and in rift zones.

Over the following months, Curiosity continued to trundle around the base of Mount Sharp, collecting samples, drilling here and there, analysing whatever it could see with its ovens, X-rays, spectrometers and lasers. It has found tantalising evidence for organic molecules in a rocky outcrop that seems as though it might once have been the mud at the bed of an ancient lake. If it is there, it is not a proof of life, of course, since organic molecules rain down on all the planets from space, where it is created in asteroids and in cosmic dust around stars and in nebulae.

In early 2014, MSL chief scientist John Grotzinger introduced a series of papers in *Science* which summarised Curiosity's work on rocks, dated to less than 3.7 billion years old, in its search for aqueous habitats on Mars that might have once supported life. '[We] describe the detection at Gale Crater of a system of ancient environments (including streams, lakes and groundwater networks) that would have been habitable by chemoautotrophic micro-organisms.'

All that means, in summary, is that Curiosity found a habitable environment at the base of Gale Crater. 'We knew there was water before, we don't know there's life,' says Vasavada. 'But until Curiosity went there, we didn't know that the rest of the conditions besides water [pH, radiation, oxidation, geology], all those factors have come together at this one site in Gale Crater. We know one spot now and by climbing Mount Sharp we'll find other spots as well hopefully and extend the time period we think Mars may have been habitable.'

The one remaining thing on Curiosity's checklist, at the time of writing, is convincing evidence of complex organic molecules. 'We're doing well on the water and the energy sources but we're only half done with the key ingredients part,' says Vasavada. 'We found carbon, nitrogen, oxygen, sulphur. We found very simple organic molecules and these ones we found are not the kind of thing that screams, "I'm a precursor of a cell, I'm the biological material that, give me a million years, I'll be DNA."'

The organic compounds they are looking for occur everywhere in the solar system, flying around in comets and asteroids, and their presence would, Vasavada says, make it an 'A+ bonus habitable environment'. It would mean Curiosity has found the right environmental conditions and also the ingredients (the precursor biological materials) that life likes to use.

'These results demonstrate that early Mars was habitable, but this does not mean that Mars was inhabited,' wrote Grotzinger in his introduction to the series of Curiosity and Opportunity results in *Science*. 'Even for Earth, it was a formidable challenge to prove that microbial life existed billions of years ago – a discovery that occurred almost 100 years after Darwin predicted it, through the recognition of microfossils preserved in silica. The trick was finding a material that could preserve cellular structures. A future mission could do the same for Mars if life had existed there. Curiosity can help now by aiding our understanding of how organic compounds are preserved in rocks, which, in turn, could provide guidance to narrow down where and how to find materials that could preserve fossils as well. However, it is not obvious that much organic matter, of either abiologic or biologic origin, might survive degradational rock-forming and environmental processes. Our expectations are conditioned by our understanding of Earth's earliest record of life, which is very sparse.'

Future missions will have to look for more direct signs of life than Curiosity is able (its mission was restricted to study geology to work out habitability). Looking for past or present signs of life more directly ('biosignatures') will be the job of, among others, the European Space Agency's ExoMars rover. This is part of a European–Russian collaboration consisting of several spacecraft, including a Mars orbiter, but it is the ground-based rover that will directly look for life when it arrives in 2018. Around the same size as Curiosity, the ExoMars MOMA (Mars Organic Molecule Analyser) will be the key instrument searching for the biosignatures of life past or present.

NASA is not being left behind by the European effort. Though the space agency has a small role in ExoMars, it has its efforts to find life focused on the follow-up to Curiosity – the Mars 2020 rover, scheduled to arrive at the red planet by the end of the decade. Whereas NASA's mantra for the first decade of twenty-first century planetary exploration has been to 'Follow the Water' and, on Mars, determine its history, that work will be subsumed in the next set of missions into a more complex, but also more direct in terms of biology, motto to 'Seek Signs for Life'.

The Mars 2020 rover will look identical to Curiosity and use the same basic platforms to get to Mars and land. But it will carry around half the kit, in terms of weight, focusing on scientific projects aimed squarely at looking for hints of biology and preparing the ground for human missions. As well as 3D cameras and a chemical analysis instrument that will be able to detect the presence of organic compounds from a distance, Mars 2020 will fly a device called MOXIE, which will try to process the carbon dioxide from the thin Martian atmosphere into oxygen. If it works it could be a way to sustain people and also make rocket fuel. Being self-sustaining on Mars (even if it makes just a small dent) could be useful in reducing the amount of supplies (and hence weight) astronauts will need to take with them from Earth. And before that, NASA wants to send another lander to Mars in 2016, called InSight, to take a look at the deep interior of the planet.

NASA also has bold plans to work out a way to get samples back from Mars, using Mars 2020 to store rocks in a box and then somehow, in a mission that is yet to be worked out, bringing that box back. 'The labs on Earth are much more capable of finding really faint signs of life than anything we could send to Mars,' says Vasavada. 'From what I hear from the biologists and palaeontologists advising us, you take a 3-billion-year-old rock on Earth and it is still very difficult to prove that there was life on Earth, even though we know 3 billion years ago there was tonnes on Earth. Unlike dinosaurs, which were yesterday geologically speaking, when you talk 3 billion years ago you're talking about little bacteria and they don't leave fossils you can put in a museum. You're looking for extremely faint chemical and textural evidence that there was a microbial community living in that rock. That's very difficult to do, even in the best labs, on rocks from Earth.'

ELEVEN

Solar System Moons

The largest planet in our solar system, Jupiter, is easy to see with the naked eye. If you are in the northern hemisphere, it is one of the brightest spots in the spring sky, rising just before midnight in the south west. Train a pair of binoculars at it and you will be able to make out its brown stripes and, if the planet is facing the right way, its giant red spot. Through the lenses, you will also see the four Galilean moons, so-called because they were first discovered by Galileo Galilei in 1610. The second closest to Jupiter is Europa. If you happen to be reading this at night, go and take a look at it now. That white, sparkling dot is our best chance of finding organisms living, at this very moment, outside the Earth.

Europa is an ice-encased moon of Jupiter, -145°C at its surface, about the size of our own Moon, 788 million kilometres (490 million miles) from the Sun and the smallest of the four satellites of the gas giant planet (the others being Io, Ganymede and Callisto) that you can see easily with a telescope.

Observations from big telescopes on the ground in the 1950s gave the first hints that there was ice or snow on the surface of Europa. Working at the McDonald Observatory in the Davis Mountains in West Texas, astronomer Gerard Kuiper saw the tell-tale signature of water in the spectra of the reflected light from the Galilean moons. Further evidence to back up his discovery came in the following decades and, by the early 1970s, the icy surface was established fact and work on Jupiter and its moons had been invigorated.

The idea that these moons might contain liquid water began in 1971 with John S. Lewis, based at the departments of chemistry and astronomy at the Massachusetts Institute of Technology (MIT). That year, he published theoretical models of the structures of the moons of the solar system, in which he showed how ice on the Galilean satellites of Jupiter and the larger satellites of Saturn, Uranus and Neptune could be melted by heat from the radioactive decay of heavy elements in the rocky cores of the moons. The moons, he concluded, would 'likely have extensively melted interiors, and most probably contain a core of hydrous silicates, an extensive mantle of ammonia-rich liquid water, and a relatively thin crust of ices'.

His ideas were further developed over subsequent years, using measurements from the flybys of the Pioneer 10 and 11 probes and more refined mathematical models of how Jupiter's gravitational field affected the planet's moons. In a 1976 paper by Lewis and fellow MIT astronomer Guy Consolmagno, the pair concluded from their computer models that Europa could have an ocean of liquid water that was 100km deep, sitting below a crust of ice that was 70km thick.

The Galileo probe passed by Europa in 1996 and sent back detailed images of the lineae – the dark streaks that criss-cross the surface of the moon, one of its most striking features. The largest of these bands was up to 20km across and they are likely to be the result of fractures in the surface where warmer ice has welled up from below to fill the gaps caused by the moon flexing in the gravitational field of Jupiter.

Galileo's magnetometer also gathered further evidence of an ocean. The probe measured an induced magnetic field on Europa through its interaction with Jupiter, implying that there was a conducting layer under the ice, likely to be salty water. By the turn of the twenty-first century, scientists were convinced that the thick Europan ice concealed an ocean that contained twice the volume of the Earth's seas.

This water has never been seen or measured directly. It is thought to be kept liquid by tidal forces acting on Europa, in which Jupiter's gravity pushes and pulls the moon out of its natural shape during its orbit. That flexing would generate enough heat to maintain a layer of salty water under the ice.

Could such an ocean be a place where life exists? Strikingly similar habitats exist on Earth and life thrives there. Both in the Arctic and Antarctic, sea ice diatoms can carry out photosynthesis while under ice cover. Mats of microbes have been found in polar lakes that are covered year-round in several metres of ice. The amount of light that gets through this ice and onto the living organisms is small – about 1 per cent of the incident light on the surface – but it seems enough on which to survive.

An alternative is to look deeper on Europa, at the interface between the water and the heated rocks at the bottom of its ocean. On Earth, hydrothermal vents spew out boiling-hot water and minerals in the pitch black. In these places, all sorts of complex organisms – giant tube worms, limpets and clams – survive by consuming extremophiles that can use the abundant sulphur-based compounds to produce organic matter without the need for light, a process known as chemosynthesis.

These are exciting, hopeful ideas. The only way we can know for sure is to send probes to Europa, to take close-up measurements or even land on its surface and drill through the ice to sample the water underneath. Both NASA and the European Space Agency (ESA) want to send missions there. 'The most interesting thing would be the chemical analysis, trying to look at possible biomarkers – complex organic materials. If there were bacteria there we would be able to detect it without going underneath,' says Bonnie Buratti, a senior planetary scientist at NASA's Jet Propulsion Lab in California. 'If there were remnants of bacteria we should be able to find them at the lineaments on Europa. There might be active plumes, there might be deposits.'

It won't be easy. Europa is deep within the magnetic field of Jupiter and the high-energy charged particles there are not too spacecraft-friendly. ESA's plan is to launch the Juice (Jupiter Icy Moons Explorer) satellite in 2022, arriving at the gas giant in 2030. It will spend at least three years making observations of the planet and its moons Ganymede, Callisto and Europa and carry cameras, magnetometers, spectrometers, a laser altimeter and an ice-penetrating radar.

In its strategy for the next decade, NASA has a relatively low-cost $2 billion mission concept called the Europa Clipper. This radiation-hardened craft would perform several dozen close flybys of the icy moon – getting as near as 25km from the surface – in long, looping orbits around Jupiter. The mission is yet to be approved but would include similar equipment to ESA's Juice mission: radar to penetrate the frozen surface and determine how thick it is, a mass spectrometer to analyse the wispy atmosphere and a high-resolution camera to take pictures of the moon's topography.

The dream mission, which is in no one's plans yet and as such would probably be just fine for Arthur C. Clarke's mysterious monolith, is to send a lander to Europa, melt through the ice to sample the water underneath and bring it back to Earth for analysis. Way beyond anyone but the most imaginative astrobiologist is the idea of sending a robotic submarine to swim the oceans, which could send back data like the machines that explore the depths of our own oceans at home.

But all of this is a tough ask. Doing that without contamination even on Earth has proved to be a tough job, as Russian and British scientists have found when trying to drill though the ice cap in Antarctica to get to the underground pools of water at Lake Vostok and Lake Ellsworth respectively.

'Trying to get something out there in space to do that', says Buratti, 'is 100 years off.' Given that it might be our only direct way to test if there is life right now on another body in our solar system, it seems like an awfully long time to wait.

To find the source of planetary scientists' greatest excitement in the water story in recent years, you have to go a little further out in the solar system, to a tiny moon of Saturn, only 500km across, named Enceladus. In 2005, the NASA spacecraft Cassini discovered a jet of salty ice crystals streaming from the moon's southern pole, which was soon nicknamed 'Cold Faithful' by planetary scientists in reference to the Old Faithful geyser in Yellowstone National Park in Wyoming. The water came out of dark fractures in the moon's icy surface, so-called 'tiger stripes'. Enceladus was not thought to be an active place, so where was the energy to produce this water coming from?

Enceladus was discovered by William Herschel in 1789, and is the sixth largest of Saturn's moons. It orbits Saturn in the densest part of the planet's E ring, a tenuous collection of fine particles, and telescopic observations in the early 1980s found that it was covered in almost pure water ice. Images from the Voyager spacecraft around the same time confirmed its size (500km across), that it was the brightest object in the solar system because its surface reflected almost all the sunlight that fell onto it, and that its surface had large smooth areas unmarked by craters.

These observations were enigmatic for a place that was thought to be geologically dead. Ice and snow tends to become dark quickly once it is exposed – think of the dirty slush on your streets a few days after a beautiful white snowfall – so the brightness of the moon's surface made no sense. Nor did its relative smoothness. Over the age of the solar system, every major object in the solar system has been smashed into by countless asteroids and comets and, unless there is an atmosphere to weather them away, rocky bodies tend to bear craters as permanent scars of these impacts.

'When we went by in Voyager, it looked like there was snow on there or something,' says Buratti, who was working on the Voyager mission in the 1980s and later on images from the Cassini probe, which visited the moon in 2005. That clean, smooth surface on Enceladus implied that the snow or ice had not been sitting there for aeons but was fresh, somehow replenished on a regular basis. At the time, Bonnie Buratti was a graduate student at Cornell University and she and her colleagues suspected that Enceladus was ejecting material from within itself, which then rained down onto the surface. 'If you have fresh ice that means there's something active going on. This was my feeling about Enceladus, that there was some type of active geology, a geyser or a volcano, and it was covering the surface.'

These ice geysers could also explain the source of Saturn's E ring, whose material was long suspected to come from Enceladus, though no one had been able to work out how. Scientists knew that the tiny particles in the ring must be constantly replenished, because they would otherwise all have been destroyed in timescales much shorter than the age of the solar system.

It took twenty-five years for these hypotheses to become

established fact. The Cassini probe spent seven years travelling from Earth and entered Saturn's orbit in July 2004. Just under a year later, in February and March 2005, the probe made its first flybys of Enceladus, taking measurements and pictures from 1,200km above the surface. During those passes, the on-board magnetometer detected a thin atmosphere around the moon, thanks to its effects on Saturn's magnetic field – charged particles leaking from Enceladus were pushing against, and distorting, the gas giant's magnetic field lines. The probe's imaging camera also saw what looked like plumes coming out of the moon, but the scientists observing it were cautious about over-interpreting what they were seeing in their instruments at that stage. 'The imaging team wanted to be careful that it wasn't some artefact, some kind of stray light in the camera,' says Buratti. 'We were really arguing about this.'

The science team requested a closer flyby of the moon, around 100km from the surface, to work out whether or not the plumes were real. Going down so close was risky, especially so early in the mission. Any errors might have sent the Cassini probe crashing into the surface or careering off into space. But their wish was granted in July 2005, when Cassini made its third and closest approach to Enceladus, this time at a distance of around 168km over the southern pole. The infra-red instruments recorded hot spots in the region and the cameras revealed a landscape littered with house-sized ice boulders and almost entirely free from impact craters. There were folded mountain ridges and cracked white ice plains streaked with a dark-green organic material – the tiger stripes.

There was also unequivocal evidence for a plume of water vapour and ice emerging from the south pole area. From warm vents near the stripes, water and dust were being ejected at supersonic speeds and, in the July flyby, Cassini flew right through one of these huge geysers.

'A lot of people who aren't scientists think that discoveries are just kind of like, you see it and you're amazed – but usually it builds up,' says Buratti about the months around the discovery of the plumes. Scientists are, by nature, sceptics and it took multiple lines of evidence to convince the majority of them about what they were seeing from Cassini. The mass spectrometer and cosmic dust detector both saw particles. The magnetometer saw a deflection of Saturn's magnetic field. The imaging camera saw plumes. The infrared camera detected a hotspot. 'All these pieces of evidence just fell into place,' says Buratti.

A few days after the July flyby, the Cassini team were convinced the plumes were real and announced their results to the world. A year later, their peer-reviewed results were published in the journal *Science*.

More than 100 geysers have now been identified on the surface of the moon and the contents include sodium chloride crystals, carbon dioxide and light hydrocarbons, as well as ice. The micrometer-sized ice particles shoot out at 60m per second and reach several tens of kilometres from the moon's surface before most of them fall back to the ground. A small amount – around 1 per cent – escapes into space, much of it forming Saturn's E ring and the rest enriching the Saturnian system with water ice.

The ice plumes raised an important question. To produce them requires liquid water to be ejected from the surface and to quickly freeze into ice as the droplets hit the vacuum of space. Scientists knew there was ice on the moon, lots of it, but where was the energy coming from to melt it into liquid? Especially on a body that was so small, cold and remote from the Sun?

Cassini scientists quickly rejected alternative sources for the plumes – one idea was that the particles in the plumes were being produced or blown off the surface – because their evidence pointed to the plumes coming from pockets of liquid water that were above freezing point, 0°C, near the moon's ice surface. At the time it was not proven, of course, merely a strong hypothesis. If it were true, though, if it were proved that there was liquid water on Enceladus, it would significantly widen the diversity of places in the solar system where it could exist. And that had big implications for the places where we might find life.

Confirmation of that liquid water came almost a decade after Cassini arrived at Saturn. In 2014, scientists led by Luciano Iess at Sapienza Università di Roma confirmed that the oceans existed under some 30 to 40km of ice that sat at the surface of Enceladus, after studying gravity measurements taken by the several flybys of the moon by Cassini over the previous decade. These flights had brought the spacecraft within 100km of the moon's surface and shown that the northern and southern hemispheres had an asymmetric gravity field. The southern hemisphere, in particular, did not seem to have enough mass at the surface to account for its gravity field and Iess proposed that something dense – liquid water – was compensating.

Those gravity measurements were made as Cassini flew past Enceladus and the moon's gravitational tug altered the spacecraft's trajectory ever so slightly. These flight alterations had a measurable effect on the microwave carrier signal being tracked by NASA scientists back on Earth and that, in turn, allowed them to work out the moon's gravity field and the distribution of its mass. At the southern pole, the scientists measured less mass than they had expected, given that they were looking at what appeared to be a uniformly spherical object. Astronomers already knew that south pole region had a depression in its surface, and therefore the resulting reduced gravitational tug was not unexpected. But the gravity field was not reduced as much as they had calculated and their conclusion was that there was an unseen, high-density feature lurking under the ice, offsetting the lack of visible mass. The best candidate for that was liquid water, which is denser than ice and also fitted with the visible jets of water found a decade earlier.

That 10km-thick layer of sea, as deep as the deepest parts of the Earth's Mariana Trench, would extended halfway or more to the equator in every direction from the planet's south pole, making it potentially larger than Lake Superior in the US. How a moon that is 1.5 billion kilometres from the Sun can have such vast oceans raises a number of questions – how long has the sea existed and is it likely to last for much longer? In other words, how does a moon with surface temperatures that can dip to -200°C and ostensibly has little else to keep it warm manage to have liquids under its surface?

The favoured idea is that the moon is being tidally heated by its eccentric orbit around Saturn. As it makes its way around, the gas giant's enormous gravitational field pulls and pushes the shape of the moon's rocky core, generating heat that keeps the water liquid. It seems that some of that water manages to push upwards and flow out of the cracks in the thick ice at the surface, emerging as one of the more than 100 plumes seen coming from the tiger stripes on the moon.

The moon's core is thought to be around 400km in diameter and the ocean itself sits directly on this silicate rock, instead of ice, as shown by the sodium chloride particles in the plume. If the plume's ice had come directly from the moon's surface, the water would have been more pure. The only way to get salts into the water is if there is a liquid reservoir interacting with and dissolving some of the moon's mantle.

That rock–water interaction leads to an intriguing possibility – there might be complex chemical reactions going on here that could produce conditions very similar to those on the early Earth, reactions that helped to create the earliest microbial life.

As we have seen, some of the best ideas for the start of life on Earth involve the formation of early cell-like life forms deep in the oceans, around hydrothermal vents, where boiling hot water spills out, powered by heat emerging from underneath the Earth's crust. These vents were a source of energy and nutrients for the emerging life and an ocean on Enceladus, with its warm rocky core, could allow for something analogous to happen there.

Along with the hydrocarbons detected in the plumes, that means any potential organisms there would have access to food, heat and water. 'It has all the elements for life,' says Buratti. 'We haven't found life there but it is a habitable environment.'

And that's what makes this so interesting: before Cassini's discovery of the Cold Faithful geyser, no one would have thought that Enceladus could become a candidate for extraterrestrial life. This tiny ice moon, so far from the Sun, is tantalising evidence that the so-called habitable zone for life (in other words, the limited region in a solar system based on distance from a star, where a planet can support liquid water and, perhaps, life) is much larger and more varied than had previously been thought.

Buratti adds that the diversity of the solar system's moons has been a continuing surprise to scientists. Before the deep space probes went out into the solar system, people expected moons just to be bland, geologically frozen in time and dead. 'We didn't think there'd even be craters there, these billiard balls out in space,' she says. That began to change when Pioneer went out to Jupiter in 1973 and saw what looked liked craters on some of the moons there. It hadn't occurred to planetary scientists then that the ice could be so strong, something like a rock-like mineral.

Understanding Enceladus and other moons has given

us, says Buratti, a 'dazzling array of habitable worlds' in our solar system and, by implication, in other star systems in our galaxy. 'Water is a *sine qua non* for life, the thing without which there is nothing. NASA said the search for life is the search for water – it's the fountain of life. There's nothing that can exist without it, that we know of.'

Jupiter's moon Europa is not out of the game yet, however. As if to get in on the act, astronomers reported in 2013 that the Hubble Space Telescope had spotted a possible plume of water coming from the moon's south pole, similar to those seen coming from Enceladus. But it was only one observation and it has never been seen since.

We have one more stop in our search for the signs of life through the solar system. Other than the Earth, Titan is the only other place that we know of which has beaches, places where sea and land come together. The liquid on the surface of Titan is not water, however, but a mixture of the hydro-carbon compounds ethane and methane. Saturn's largest moon is as big as Mercury with an atmosphere composed mainly of nitrogen, and a small amount of methane and hydrogen, that is ten times more massive than the Earth's. The pressure on the ground is 50 per cent higher than here and the surface temperature is around 9K (-179°C).

When Voyager 2 approached Titan in 1981, its sensors could not penetrate the opaque red-orange haze in the air but found that the upper atmosphere was rich in sunlight-induced organic chemical reactions that produced copious hydrocar-bons and nitorgen-carbon compounds called nitriles, a poten-tially rich stew of materials that could be used by a life form to grow. There is water here but, with such a cold surface temperature, there is little chance that it would be liquid.

'Titan has, I believe, been put in our solar system to test our faith in water. Here is this moon of Saturn that looks, for all the world, like the Earth,' says Lynn Rothschild, an astrobiologist at NASA's Ames Research Center in Moffett Field, California. 'It's got lakes and mountains, the problem is that it's so cold that the mountains are made of frozen water and a lot of the boulders and the lakes are liquid ethane and methane. There's lots of organic matter. It would be a wonderful place to have life.'

What a Titanic life form would have to rely on for a solvent would be the ubiquitous liquid methane and ethane that flows on the surface. What might life look like, then, if it could survive in a world like this, in such a different medium to any life we know?

At first glance, methane (or ethane) would be a horrible choice of solvent. It is an organic molecule with no polarity, meaning very few things will dissolve in it. It is also extremely cold, as befits its presence on Titan.

But such limitations do not preclude the possibility of life – we just need to get more creative in how we think life works. In 2005, the European Space Agency's Huygens probe landed on Titan. Named after the Dutch astronomer Christiaan Huygens who discovered Titan in 1655, the probe was part of the international Cassini mission to Saturn. Huygens was dropped from its mothership on Christmas Day in 2004 and reached the surface of Titan a few weeks later, landing in what seemed like mud near a region called Xanadu. For 90 minutes, the 300kg probe sent back copious measurements and images of an orange surface covered in clouds of methane and rocks of water ice.

On the day Huygens landed on Titan, NASA's Chris McKay submitted a paper to the journal *Icarus*, in which he hypothesised how we might go about spotting a life form on Titan that would have a radically different biochemistry to our own. If you ignore water, the search for life usually rests on hunting for biomolecules such as amino acids. Of course, that won't work for Titan because we have no guide from Earth on how the biology or biochemistry of the hypothetical organism itself might work and, therefore, what its significant biomolecules might look like.

McKay decided to bypass the unknown biochemistry of the organism itself and looked for a more fundamental process of life that he assumed would be ubiquitous. He settled on its pollution of the environment. 'We call it pollution when humans do it but you can use the same word to refer to what cyanobacteria did to make the oxygen that's in the Earth's environment. Or what trees do every spring when they soak up carbon dioxide and every fall when they release carbon dioxide.'

Biology interacts with its environment and, if the biology is widespread, the effect is widespread too. And that effect can be detected in the ratio of gases in the atmosphere of a planet. The water-based life on Earth has global impacts because water is found everywhere on Earth; similarly, methane-based life on Titan would have global effects because that chemical is found everywhere on the moon.

But what would you look for? By analysing the various chemical reactions that would be available to Titanic life as a reliable source of power, McKay concluded that a methane-based life form would most likely want to consume hydrogen, in the way Earth life consumes oxygen.

When you eat something, your body takes in the organic material, breaks it down into simpler hydrocarbon molecules (such as glucose) and then reacts it with the oxygen you breathe in order to release chemical energy. That metabolism – oxidising organic material – is common throughout life on Earth and the waste products are water and carbon dioxide, which come out when you exhale.

On Titan, the favourable (and most widely available) chemical reaction to power metabolism would involve hydrogenating the organic material that an organism eats.

That would provide energy comparable to the energy that organisms on Earth get from their metabolism. 'We predicted that the best food on Titan would be the acetylene in the atmosphere plus the hydrogen in the atmosphere. An organism sitting on the ground on Titan, if it could eat acetylene and hydrogen, it would be deriving an energy source that is very favourable compared to biochemical energy as we see on Earth.'

The waste products of the Titanic metabolism would include methane. 'We wouldn't see methane because the planet is already flooded with that so we didn't predict methane as a biosignature because there's a big background signal,' says McKay. 'But hydrogen is at trace levels [on Titan] and, if the organisms were consuming it, we would be able to see a depletion.'

The signal to look for, then, is an anomalous depletion in hydrogen near the surface of the planet because the organisms are eating it up. There are no other natural chemical sinks for hydrogen expected near the surface.

McKay's Titanic organism would look nothing like you can imagine. In fact, to our eyes it would barely be living at all. Life on Earth has evolved to live in water, which is, chemically, a very strong and aggressive solvent. With water, life gets something of a free ride, according to McKay – micro-organisms on Earth are small and live in 'thick, rich solutions', able to rely on water's chemical anomalies and strengths to diffuse nutrients and waste in and out of their cells without much effort on their part.

That won't work on Titan. The liquid methane medium is thin, cold and not very active. Unlike the small spheres on Earth, cells on Titan (if you can call them such a thing) will need a huge surface area to volume ratio. 'On Titan, imagine like a giant sheet of paper, very, very thin; huge surface area and membranes that actively pump and move nutrients. They'll have more control over the interactions with the environment rather than letting the properties of the solution that they're in dictate their interactions.'

On Titan, the raw materials are less concentrated; they're less reactive and not moving very fast. It would be like the difference between living on a tropical island where there is fruit in the trees and all you have to do is lie back and let it fall on you, versus living in the frozen Arctic where you would need to hunt for every meal. The organisms on Titan are going to be working for a living, far harder than on Earth where the water provides much richer, thicker, active soup and, in the higher temperature, things move faster.

Recognising these Titanic films as life would be difficult even if we could somehow get a sample. The standard test to work out what something is and whether it is alive is to try to grow it in a well-understood culture medium in a laboratory. Which would no doubt be hard to do for something as strange as the Titanic film. In any case, given its evolution in extreme cold, it would probably grow too slowly for us to recognise anything on human timescales. We could try looking for biomolecules such as amino acids or DNA, but again the biochemistry might be so alien that we just don't recognise any of the molecules as signifiers of life.

'That's why we have to take the approach of what does it do to its environment,' says McKay. Defining life is not straightforward even on Earth (is it something that reproduces? Something that grows? Does both? Something else? We will look at this more closely in the next chapter) but one of the more general properties of living things is that they manage energy and materials in a way that would not happen without them.

On Earth, a spoon of sugar would sit happily exposed to the oxygen in the air for as long as the lifetime of the planet, without reacting. Yet a reaction between these two elements goes on inside living cells every day; it is the basic process upon which all of life here is based. Life pumps chemicals and energy in and out of the environment in a way that chemistry cannot by itself because life has evolved enzymes, biological catalysts, short-circuit chemical reactions that would, otherwise, be energetically inhibited. To get sugar to react with oxygen, thereby releasing its chemical energy, you need to apply a little energy to the sugar, in the form of heat, to get it started. Enzymes allow Earth's biology to do this without actually burning anything.

On Titan, the acetylene–hydrogen reaction is energetically inhibited. Same as our sugar–oxygen reaction, the Titanic ingredients could just sit next to each other through to the end of the universe without reacting, even though there is enormous energy to be released by doing so. If our sample of potential Titanic life were somehow catalysing the acetylene–hydrogen reaction, though, then we would make a good bet that it was alive. 'The stranger life gets, the more we have to rely on its general, rather than specific, properties,' says McKay.

Mckay wrote his paper in 2005, the same year that the Huygens probe landed on Titan and took measurements on the moon. Six years later, a team at Johns Hopkins University in Baltimore, Maryland, used Huygens' data to calculate that there might well be a measurable depletion of hydrogen at the surface of the moon. They referenced McKay's prediction that this would happen if methane-dwelling life were present.

'For a while I was extremely excited,' says McKay when he heard about the work. 'I thought, "Oh my God, this is amazing, not only does it detect life but we predicted it."'

On further reading, though, he realised he had some time to wait before his Titanic prediction could be proved right or wrong. The Johns Hopkins conclusion was based on the result of a computer model result, rather than being a direct observation. 'They did a model of hydrogen in the atmosphere on Titan and they found that they could only make their model match their observations if they postulated a sink at the surface,' he says.

Though it got the astrobiology community focused on the hydrogen sink as a potential biomarker, the only way to truly confirm McKay's hypothesis is to send another probe to measure the hydrogen flux at the surface of Titan. The Huygens probe, in principle, could have done that but its instruments were not optimised for that task.

'What you'd really like is measurements of hydrogen over time and over latitude,' says McKay. 'What you'd predict is a stronger depletion over places where there's liquid, in the northern hemisphere, less in the southern hemisphere. And there might also be a seasonal dependence. You would treat it like you would carbon dioxide on Earth. When you track that on Earth you find there's a hemispheric asymmetry and a seasonal variation, all of which traces back to biology.'

McKay proposed the Titan life hypothesis partly as a thought experiment but partly also as an extreme test to the question of whether life forms can survive without a solvent as useful as water is on Earth. If life forms can find a way to use something so different, so cold, non-polar, something that is such a poor solvent, then that would open up a whole slew of other possibilities for potential solvents for life.

Water is the material used by life on Earth but what if life in the rest of the universe has been more creative? We will come back to this idea in the final chapter but, before we do, we should expand our search for water beyond our known solar system.

TWELVE

Beyond the Solar System

The search for life and water in space is, right now, a search for habitability. In our discussion of the various environments in the solar system that contain liquid water, we have been skirting around a topic that has been at the forefront of scientific discussion for several decades but which has quickened in pace over the past ten years as the search for life begins to leave the solar system. It is worth tackling that topic head on – what is a habitable zone?

We used to talk about the habitable zone around a star as a distance from a star that was not too close (so that liquid water all evaporated) and not too far (so that water only existed as ice). The exact distances will depend on the star and its energy output, of course. Earth is the only planet within the habitable zone of our Sun.

Astronomer Rami Rekola sums up the traditional, Goldilocks-inspired thinking when he writes: 'The pressure–temperature range for liquid water may be found in places with sufficient, but not excessive heating, and some, but not too dense an atmosphere. Interstellar gas clouds are too sparse and cold, stellar atmospheres too hot, bodies orbiting stars very close are too hot, and bodies very far from stars too cold – unless they have a local source of heat. Bodies that are too small cannot retain any atmosphere – excluding asteroids, comets, small moons and small planets. As a conclusion the habitable zones are formed on moons and planets at suitable distance from a mother star and with permanent atmospheres.'

Those basic constraints have also been part of the guidelines in our search beyond our own solar system. As we find more exoplanets, planets that orbit other stars, we have tended to look in regions that we might define as the 'Goldilocks zone' around those stars.

But these constraints have steadily been eroding away. The prediction and subsequent discovery of liquid water on moons in the outer solar system – Europa and Enceladus among them – was one factor, though they were partly accounted for in Rekola's definition. Their liquid water exists thanks to the tidal forces they experience as a result of moving around in the intense gravitational fields of their parent planets. At any point in their orbit, the gravitational pull on one of their sides is bigger than that on the other, causing the moon to be pulled out of shape. As they travel around their parent body, those gravitational forces flex the moon's shape and cause it to heat up. The ice on the moons, which should be completely solid given their distance from the Sun, partly melts.

'Long ago, we introduced two terms – one is the radiatively heated habitable zone, which is where Earth is, and the other is the tidally heated habitable zone, which is what Europa and Enceladus are in,' says astrobioloist Chris McKay of the NASA Ames Research Center in Moffett Field. 'They are very different and the prospects for life should be considered for both. But right now we have no way to detect these objects in the tidally heated habitable zones [outside our solar system]. We can't detect Europas and Enceladuses around the giant planets around other stars.' This is a problem for the exoplanet search, he adds, because it means we will miss lots of planets in our quest. 'Do we think we've got them all? We didn't predict [liquid water on] Europa and Enceladus; we stumbled across them. Maybe there's more ways that there could be water on worlds that we haven't even thought about.'

The traditional definition of the habitable zone also implies looking for liquid water around Sun-like stars and nowadays even that looks too restrictive.

Astronomers used to think that stars whose masses were much greater or smaller than our Sun would be inhospitable to biology and particularly to the evolution of complex or intelligent life. Because their nuclear reactions happen at an increased rate, the most massive stars burn through their hydrogen fuel too fast, over just a few million years, and this is not long enough for the multi-billion-year timescales that, on Earth, allowed the emergence of the life we see. Low-mass stars, on the other hand, burn slowly and their energy output was thought so feeble that, for an orbiting planet to have liquid water, it would need to be right up close. In such a tight orbit, the planet would become tidally locked to the star, meaning it would keep the same side facing it all the time – in our solar system this happens between the Earth and the Moon. These planets could not support liquid water, it was thought, because a tidally locked planet's atmosphere would boil off on the side facing the star and freeze on the dark side.

That thinking has been changing, thanks to several factors. The increasing discovery of life in unexpected places in extreme habitats on Earth such as hydrothermal vents in the dark of the bottom of the oceans has given us fresh ideas about the potential limits of living things. The discovery of liquid water in unexpected places in our solar system has given us potential new habitats. And our ability to spot star systems outside our own solar system widens the pool of potential life-supporting worlds. Some astronomers go so far as to argue that certain low-mass stars, known as 'M dwarf' stars and which make up some 75 per cent or more of the stars we see and account for half of the stellar mass of the Milky Way, might be the best and most likely locations for habitable planets that can stay habitable for many billions of years. Those planets may not be blue and have oceans at their surface like Earth, but they are at least worthy of consideration nonetheless.

Stars are classified on an alphabetical scale from hottest and most massive to the coolest – O, B, A, F, G, K, M and L. Each letter also has a fractional division from 0 to 9 that goes from hottest to coldest in that class. By this classification system, our Sun is a dwarf G2 star. The biggest stars are the hottest and burn out fast, whereas M dwarf stars can remain stable for a very long time – theoretically between 50 billion years for the biggest M-class stars to trillions of years for the smallest. So far in the history of the Milky Way, no M dwarf star has had enough time to come to the end of its life. Due to their low temperatures, the light from these stars glows red and they tend to be smaller than our Sun.

In a review that reconsidered M-class stars for potential habitability, a group of astronomers and biologists laid out the advantages that such a system could have. 'Based on a simple energy-balance model, it was argued that atmospheric heat transport could prevent freeze-out on the dark side,' they wrote in the journal *Astrobiology*. 'More sophisticated three-dimensional climate modelling suggested that a surface pressure of as little as 0.1 bar could prevent atmospheric collapse on the dark side, while a surface pressure of 1–2 bars could allow liquid water on most parts of the surface.'

They conceded that the magnetic activity and radiation flares from the M dwarfs would be a problem, because of the large amounts of life-damaging ultraviolet-B rays that would arrive at any closely orbiting planet. But this could be mitigated for anything living on the surface if there were geological or atmospheric processes that, like on Earth, created ozone and protected the fragile life at the surface.

And this is all assuming the worst-case scenario for these M-class star systems. Not all close-orbiting planets get tidally locked and if liquid water can move freely around the planet, it can redistribute the heat so that the temperature differences across the surface are not so pronounced. The specific geophysical conditions on the planet could also aid with habitability. Even if a layer of ice were to form on the dark side of the planet, say, if it was geologically active and had Earth-like levels of geothermal heat coming from its interior, that heat could keep some of the water liquid. Even better, if the ocean basins on the dark and light sides were connected, it could maintain 'a vigorous hydrological cycle' according to the scientists.

The limits of the habitable zone (HZ) around an M star, concluded the scientists in *Astrobiology*, could be defined by similar processes to those that define the limits around a G star like ours. 'The atmospheres of planets in the HZ of M stars would certainly be very unfamiliar to us in terms of their circulation, radiation, and chemistry,' they wrote. 'However, if planets can keep their atmospheres and water inventories, none of these differences presents a large obstacle to their potential habitability. Therefore, from the point of view of atmospheric and climate science, planets in the HZ of M dwarfs should have almost as high a chance of being habitable as planets in the HZ of G stars.'

The casting net for life, then, keeps getting wider. Rekola writes that the main constituents of life – water and organic molecules – are known to be widespread in the universe. 'Interstellar clouds harbour a vast variety of organic molecules, tens of which – including some amino acids – have been identified by radio astronomers. Meteorites, comets and asteroids have been found to contain a multitude of organic material. The distribution of heavy elements is adequately universal to suggest that most stellar systems have the same composition and same chemical possibility for emergence of life.'

With an increasing number of places now believed to contain material that might be used for nascent life, it is not hard to be optimistic. Rekola concludes: 'It is hard to imagine life as something special or rare, considering how much there is of it on Earth and how well it adapts to different conditions. It is likewise hard to believe there would not be an abundance of habitable environments in the Milky Way and, most likely, in most galaxies.'

Kepler-186 is a red dwarf star, technically known as an 'M dwarf', that is half the size and mass of our Sun and situated around 500 light years away in the constellation Cygnus. There are five planets in orbit and one of them, Kepler-186f, is the spitting image of Earth – within 10 per cent of the same size and probably rocky. Most important, it is within the most traditional definition of the habitable zone of its parent star. There was some excitement when it was discovered in April 2014, because Kepler-186f became the very first Earth-like planet to be found within the habitable zone of a star.

Before Kepler-186f, all exoplanets found within habitable zones had been at least 40 per cent larger and the small number of Earth-sized rocky planets that had been spotted had been too close or too far from their stars to be thought capable of life.

Kepler-20e, for example, was found in the constellation Lyra, 1,000 light years from Earth, orbiting its sun every six days. Before Kepler-186f came along, it was the most Earth-like found so far, thought to be made from an Earth-like blend of iron and silicate rocks and just a fraction bigger than our planet. With surface temperatures around 700°C, however, the chances of liquid water on this fiery hell were slim to zero.

A dozen or so planets have been found in the right place in relation to their stars – within the habitable zones – but planets such as Kepler-22b and Kepler-62f were all much bigger than Earth and thought to have thick atmospheres, more like the gas giants Jupiter or Neptune in our own solar system than the inner terrestrial planets, and are therefore not conducive to the life with which we might be familiar.

Kepler-186f was the first 'Goldilocks' planet, the right size and the right distance from its star, about 0.36 astronomical units (AU) from its star (one AU is the distance between the Earth and the Sun), which makes it closer in than the orbit of Mercury (0.38 AU) in our solar system. The planet orbits its star every 130 days and gets around a third of the energy from its star that Earth gets from the Sun. M dwarfs are cooler and dimmer than our Sun so, at noon, its star appears as bright to Kepler-186f as our Sun does to us about an hour before sunset. Being in the habitable zone does not mean that the planet itself is suitable for life, however – there are many more important factors to be worked out by future satellites and telescopes, such as the composition of the planet's atmosphere and whether that has the right materials and properties to allow life forms to thrive.

In the search for the possibility of life (and water) on planets other than our own, exoplanets have blown the story wide open. Suspected to exist by astronomers and thinkers for centuries, there was no way to detect exoplanets similar to Earth until the last few years of the twentieth century. The first exoplanet orbiting a star was confirmed in 1995 by Michel Mayor and Didier Queloz of the Geneva Observatory in Switzerland. They spotted a Jupiter-sized planet orbiting 0.05AU from the sun-like star 51 Pegasi, which is fifty light years from us in the constellation of Pegasus. A handful more exoplanets came along in subsequent years through painstaking observations of individual stars but the pace accelerated after the launch of the Corot and Kepler probes in 2007 and 2009 respectively, designed specifically to look for new worlds.

Exoplanets are found in several ways. NASA's Kepler probe, for example, kept a close watch on some 150,000 stars for the tell-tale signs of orbiting planets. As a planet transits past a star, the starlight reaching Kepler drops by a minuscule amount and measuring the size and frequency of this effect tells astronomers how big a planet is and how quickly it orbits its star. Another approach to finding exoplanets is to watch a star for the wobble in its position caused by the gravitational field of its orbiting objects.

These techniques are not flawless – the transit method requires the star, planet and the observing probe to be precisely aligned and will therefore miss the majority of planets in the galaxy that don't happen to line up. The gravitational method is indirect and can 'find' planets where none exist if the wobbles are misinterpreted. To guard against false positives and to ensure that the dips in light are not coming from stars orbiting (and eclipsing) each other, exoplanet candidates have often been watched for several years before they get confirmed and the evidence for their existence checked by other telescopes on the ground and in space.

That verification bottleneck was breached in March 2014, when NASA announced that it had identified its biggest haul of confirmed exoplanets so far: 715 new exoplanets orbiting in multi-planet systems around 305 stars. The vast majority of these were smaller than Neptune and four of them were less than 2.5 times the size of Earth and in their stars' habitable zones. Before that, Kepler had identified 246 exoplanets

– the 715 new ones meant that more than half of the exoplanets found in history to that point had come from one spacecraft over the previous few years.

Astronomers were able to announce so many because they had sped up their verification process with a technique called 'verification by multiplicity', which relies on that fact that, if a star shows multiple dips in light over a short period, planets must be responsible because it would be difficult for so many stars to orbit so close together and remain stable.

Thanks to these technological innovations, the rate of progress in the hunt for exoplanets has been stunning. As of summer 2014, we know of 1,800 confirmed exoplanets and there are many thousands more candidates. Most of these are gas giants, their size making them easier to detect. Increasing precision, though, has allowed astronomers to find a wider range of sizes. The smallest found so far is twice the size of our Moon, while the biggest is almost thirty times larger than Jupiter.

The astronomers' rule of thumb is that there is, on average, likely to be a planet for every one of the 100 billion stars in the Milky Way. In addition to that, there could be trillions of free-floating planets in interstellar space, untethered to any star system.

John Johnson of the California Institute of Technology was able to confirm the rule of thumb estimate by studying the orbits of the five planets around Kepler-32, an M dwarf star, and estimating how typical that system was compared to others. Planetary systems, he found, would be the cosmic norm, most of them around the cooler, smaller M dwarf stars that make up around 75 per cent of the Milky Way's population.

To focus that number more on the likelihood of finding life, Erik Petigura at the University of California, Berkeley, also trawled through the Kepler data for planets that had a radius up to double that of Earth's and which orbited within the habitable zones of their stars. He found that 22 per cent of our galaxy's sun-like stars would have rocky planets circling them, which would get around the same amount of light energy as the Earth gets from the Sun – that means a possible 2 billion planets with the potential for liquid water on their surfaces. Of the 100 billion stars in our galaxy, around 10 per cent are like the Sun and Petigura's estimate that one in five of them might host rocky planets in the habitable zone was far higher than most cautious astronomers had previously dared hope.

The explosion of ideas and discoveries has led to an undeserved ennui with exoplanets in the world outside astronomy. Finding new worlds is now such a common thing that newspapers and TV barely cover announcements unless they happen to be like Earth or within habitable zones.

Kepler-186f was found after looking through three years of data from the Kepler satellite. The system as a whole contains four other planets, Kepler-186b, Kepler-186c, Kepler-186d and Kepler-186e (there are clearly no prizes for imaginative naming in exoplanet astronomy). They are all roughly Earth-sized (mostly a bit bigger) and orbit their star every four, seven, thirteen, and twenty-two days respectively. They are all closer in than the estimated start of the habitable zone for Kepler-186, which is around 0.22AU, making them too hot for life as we currently know it.

Generally, planets in the habitable zones of M dwarf stars are easier to spot because the proportion of starlight they block during a transit is greater and these planets also tend to be closer and transit more quickly, so more transits can be recorded in a fixed period of time – Kepler-186b, c, d and e were found in the first few months of observations. It took a little longer to record enough transits of Kepler-186f in order to reach the statistical significance required to confirm the discovery.

Since the transit method measures a planet's size but not its mass, astronomers cannot be certain yet about the composition of Kepler-186f. Theoretical models of planets like this, though, show that it is unlikely to have the kind of thick atmospheres we see on Neptune and, from evidence of the Earth-sized planets in our own solar system, we know that planets of this size tend to be made mainly of rock, iron, water and ice.

Measuring the composition of planets so far away is at the very edge of astronomical technology. So far, there have been strong hints of water in the atmospheres of exoplanets including HD189733b, HD209458 and XO1b, but not all the results have been uniformly accepted and working out how much water there is with current instruments is proving difficult, given the amount of noise in the spectroscopic readings.

HD189733b, sixty-three light years away in the Vulpecula constellation (the fox), was discovered in 2005 when French astronomers saw it transiting across its parent star. It is a gas giant, some 13 per cent bigger than Jupiter, which orbits its star once every 2.2 days, travelling at a whopping 547,000kph. It is one of the most well-studied exoplanets and was the first to have its colour determined.

Astronomers at the University of Oxford pointed the Hubble Space Telescope and measured the light coming from the planet and its star and compared that to light coming just from the star to find that HD189733b would appear to be a rich navy blue to the naked eye. Though that is, tantalisingly, the colour of a deep water ocean, in this case it is more likely to be the result of clouds on the gas giant planet that are filled with silicon compounds. In essence, this planet's atmosphere is filled with twinkling drops of glass.

Detecting water on exoplanets from the ground on Earth is almost impossible because our atmosphere has so much water in it that it will contaminate any readings. Giovanna Tinetti of University College London used data from NASA's Spitzer space telescope in 2007 to show that there was water in the atmosphere of the planet HD189733b and follow-up observations with Hubble Space Telescope confirmed that finding but also showed that the amount of water was very low – one tenth to one thousandth of the amount predicted by planetary-formation theories. This has important implications for finding water on rocky planets. Nikku Madhusudhan of the University of Cambridge, who led the Hubble work on HD189733b, said that any instruments on future space telescopes might need to be more sensitive than previously thought in order to make sure they detect water and that astronomers 'should be prepared for much lower water abundances than predicted when looking at super-Earths', in other words large planets that are bigger than Earth but smaller than the gas giants such as Neptune and Uranus.

Madhusudhan's measurements also took in the planet HD209458b, which was first found in 1999 and orbits a Sun-like star that is 150 light years from us in the constellation Pegasus. That planet and also XO-1b, orbiting a yellow dwarf star around 560 light years away in the constellation Corona Borealis, are both hazy worlds with faint confirmations of water from several observations by the Hubble Space Telescope, which took on the extraordinarily difficult job of looking at the absorption spectra of the planets' atmospheres to work out what elements might be present there. Of all the planets studied so far, HD209458b had the strongest indications of water.

At this point, we have very little information about these planets. For many of them we only have as little as one pixel on a wide image taken by a telescope in space. 'Can you imagine that everything we knew about planet Earth was squashed down into one pixel? You'll get everything you can out of the spectrum but it's hardly going to give you the richness of human experience or vegetation. It's really very bare bones,' says Lynn Rothschild, an astrobiologist at NASA's Ames Research Center in Moffett Field, California.

Getting under the skin of these details will be the job of the next generation of exoplanet-hunting observatories such as NASA's Transiting Exoplanet Survey Satellite (Tess) and its James Webb Space Telescope. The European Space Agency will also get in on the act with launches of its Characterising Exoplanet Satellite (Cheops) and Planetary Transits and Oscillations of stars (Plato) probes in the coming decade, both dedicated to studying exoplanets. Between them, they will be able to look for more planets in habitable zones and also work out what is in their atmospheres. All of which will be required in the meaningful search for life.

Looking for individual elements in the spectroscopic signatures is our best bet of hunting for the signs of life elsewhere in the cosmos. Water, of course, is a key thing to look for but not a sign of life itself. Oxygen is a good potential biomarker, since a raised amount of it, compared to the amount you might expect as a result of basic planetary formation, could mean some sort of biological process is going on. According to Chris McKay, levels of oxygen over a few per cent on an exoplanet would be consistent with the presence of multicellular organisms and high levels of oxygen on Earth-like worlds might indicate oxygenic photosynthesis. On Earth, for example, oxygen concentrations became higher because of photosynthesis in plants, where it is the waste product. 'I would try to search for something that would indicate there's some interesting biology – maybe a signature of methane or a signature of the vegetation on the surface,' says McKay, echoing his hypothetical ideas for detecting a life form on Titan. On Earth one of the characteristics of the surface is the so-called 'red edge' of photosynthesis. Anyone looking at the Earth with a spectrometer would see red and infra-red light missing in the sunlight that is reflecting off our surface. This is due to the chlorophyll in plants absorbing so much of these wavelengths of light in order to do their work.

Further limits can be placed on extrasolar systems based on more general physical characteristics, again based on what we know about the limits of life on Earth. Temperature is key, writes McKay, both because of its influence on liquid water and because it can be directly estimated from orbital and climate models of exoplanetary systems.

Studying extremophiles shows us that life can grow and reproduce at temperatures as low as -15°C, and as high as 122°C. Studies of life in extreme deserts show that on a dry world, even a small amount of rain, fog, snow and even atmospheric humidity can be adequate for photosynthesis to support a small, but detectable, microbial community.

Life on Earth is capable of survival at very low levels of light, less than one ten-thousandth of the solar flux we see at the surface. Geothermal energy could also be used by organisms, the presence of which could be estimated by the bulk density of the planet. Ultraviolet or ionising radiation (which is damaging for us because it can break DNA and cause cancer) can be tolerated by many micro-organisms on Earth at very high levels and is unlikely to be life-limiting on an exoplanet. The availability of nitrogen for biological purposes, though, may limit a planet's habitability.

'Life is a planetary phenomenon,' concludes McKay. 'We see its profound influences on the surface of one planet – the Earth. Its origin, history, present reach, and global-scale inter-actions remain a mystery primarily because we have only one datum. Many questions about life await the discovery of another life form with which to compare. Mars early in its history is probably the best prospective target in the search for extraterrestrial life forms, although Europa and Enceladus are also promising candidates due to the likely presence of liquid water beneath a surface ice shell and the possibility of associated hydrothermal vent activity. In any case it is likely that our true understanding of life is to be found in the exploration of other worlds – both those with and without life forms. We have only just begun to search.'

THIRTEEN

Alternatives to Water

'There is a famous book published about 1912 by Lawrence J. Henderson ... in which Henderson concludes that life necessarily must be based on carbon and water, and have its higher forms metabolising free oxygen. I personally find this conclusion suspect, if only because Lawrence Henderson was made of carbon and water and metabolised free oxygen. Henderson had a vested interest.'

Carl Sagan, 1973

Until now, we have made the assumption that water is an absolute requirement for the discovery of life. It is certainly necessary for life on Earth, for the many and varied reasons we have already discussed. Life on Earth has been categorised into millions of species but, at the molecular level, we are all the same, made from DNA and proteins acting in solutions of water. There are many environments on our planet that are extreme by human standards but where life thrives nonetheless – boiling springs under the ocean and sub-zero regions under ice at the poles. But all of these life forms we know evolved from a single common ancestor that lived around 3.8 billion years ago. The search for life in dry environments or which does not use liquid water has so far been fruitless, though people continue to search for possibilities at the poles and in hot deserts.

Without water there would be no life (and not much else of interest) on the Earth and NASA's 'Follow the Water' strategy is a sensible starting place in the search for life elsewhere. The exploration of other worlds has been tied to an excitement and need to find water or hints of water in different parts of our galaxy.

But all of that is, as Carl Sagan might have said, a very hydrocentric view of the universe. When searching for life elsewhere, we have looked for life like that we see on Earth, based on water and the biomolecules – nucleic acids, proteins and sugars – that work with it. And you can understand why: we have one complex, refined example of how biochemistry works and it is difficult to imagine others that could approach its flexibility and power. This, biologist William Bains argues, is chemical parochialism.

The universe is a big place and, just because our only confirmed example of life uses water, we would be illogical in dismissing other possibilities. If nothing else, astrobiologist Chris McKay's example of a theoretical life form on Titan shows us that our water-obsessed direction could lead us to become blind to other possibilities for the emergence of life on other planets (and, indeed, alternative forms of life on Earth).

There is no chemical reason that the various functions we know are needed for life could not operate with other systems of molecules, alien to ours. 'It is the nature of the liquid in which life evolves that defines the most appropriate chemistry,' writes Bains. 'Fluids other than water could be abundant on a cosmic scale and could therefore be an environment in which non-terrestrial biochemistry could evolve.'

But what about the myriad weird properties of water, which are so crucial to life here and which are impossible to mimic with other solvents? We need to be careful with cause and effect here. Yes, water has a lot of unusual properties and life on Earth has moulded itself to them, but we cannot know that life everywhere intrinsically requires those unusual properties in order to function. It could be the cosmic equivalent of being amazed how your hand has evolved to exactly fit the glove you find in your drawer.

The universal reliance of Earth life on water could mean that the biochemistry we know is the only chemical way to manage the requirements of life. Or it could simply be the best chemical way to manage the requirements for life, given the conditions in which our ancestor species found themselves on the Earth. They needed a solvent and the most common liquid around happened to be water, so what we see is the best adaptation to life in water. Even more extreme is the possibility that what we observe about life on Earth tells us nothing about the best environments for life more generally – perhaps the compatibility between our life and water is neither the best or most universal way to organise things, but just reflects a shared historical accident.

Different fluids, if they supported life, would employ different chemical means to make the organisms work. Indeed, we might not even recognise their biological and chemical signatures if we were out hunting other planets and were only equipped to search for life based on water.

'Given the inevitability of human missions to Mars and other locales potentially inhabited by alien life, an understanding of the scope of life will improve researchers' chance to study such life before a human presence contaminates it or, through ignorance or inaction, destroys it,' wrote a panel of astrobiologists and other experts convened by the US National Research Council in 2007 to define the variety of possibilities for life outside Earth. They found 'no compelling reason' for life being limited to water as a solvent, even if we constrained ourselves to carbon as the backbone for the biomolecules involved.

'In water, varied molecular structures are conceivable that could (in principle) support life, but it would be sufficiently different from life on Earth that it would be overlooked by unsophisticated life-detection tools,' wrote the experts. 'Evidence suggests that Darwinian processes require water, or a solvent like water, if they are supported by organic biopolymers (such as DNA).'

But what would this non-water world look like? Tackling this question involves entering a world of speculation, and I should be clear that much of the detail we will discuss from here is hypothetical, with little hard evidence from other worlds to give us conclusions one way or the other. But the discussions are a useful way to understand life on our own planet as well as opening up the creativity of astrobiology to the possibilities on other worlds. To tackle this, it's worth breaking the problem of non-aqueous life into parts, to identify what we should be looking for.

The first part is, philosophically and scientifically, probably most tricksy – what do we mean by life?

Who hasn't asked this question in any number of contexts? At school you may have been taught to recognise life by its various properties – movement, respiration, sensitivity, growth, reproduction, excretion and nutrition (usually summarised in the mnemonic 'Mrs Gren'). The problem with this is that, within seconds of being taught this at school, pupils can usually think of non-living things that fulfil all or most of these functions. Fire, for example, moves, reproduces, grows and requires 'nutrition'. To some degree it is also sensitive to its environment and 'respires' by burning the material around it. We don't, most of us anyway, consider fire to be alive.

You might expect scientists, especially those looking for

life or signs of it elsewhere in the cosmos, to have a more precise definition. You'd be wrong. Carl Sagan, Erwin Schrödinger and many others have suggested a combination of physical, metabolic, biochemical and thermodynamic properties that, together, define something as living, more sophisticated versions of the Mrs Gren idea above. But there is no single simple answer.

Entire books and research careers are based on trying to answer the philosophical and scientific questions around how to define life and most of that is outside the scope of our discussion. We need some sort of working definition of life, though, if only so that we can all agree on what we are talking about and, later, decide what it is we are looking for in space.

Describing characteristics is not a bad place to start. In addition to the Mrs Gren list above of how life functions, we can add a few more concepts that hold true for the life we see (and the life we expect to see).

Life is chemical; it is made of molecules that undergo very specific chemical transformations (the breakdown of sugars in cells, for example, to extract energy), directed by other molecules (proteins) whose structures are inherited between successive generations. And the information to make those inherited molecules is also carried by molecules (DNA in our case).

All of these molecules interact with a solvent (water, on Earth) – some are soluble, others are not; some react with the solvent, others do not – in order to build and maintain the structures and biomolecules they need.

When the definition of life was considered by a NASA-convened panel in 1994, they came up with an elegant solution: 'Life is a self-sustained chemical system capable of undergoing Darwinian evolution.'

Darwinian evolution means evolution by natural selection, sometimes described as 'descent with modification'. This is the idea that a molecular system will tend to replicate itself imperfectly and then pass on those imperfections to its next generation. In the case of life on Earth, the replicating molecules are DNA and we call the imperfections mutations. Darwinian evolution, therefore, implies more than reproduction.

The NASA definition is good, but inevitably not complete. Life on Earth is based on DNA that contains a code for building the proteins within an organism, used in everything from its structures to its signalling. DNA passes between generations, along with mutations, and that drives Darwinian evolution.

None of that tells us how life on Earth began or whether there have been many types of life in the Earth's history – the symbiotic DNA-protein system could not have appeared, fully formed, out of the swamp. It is possible that the precursors to life based on the symbiotic relationship between nucleic acids and proteins were not capable of Darwinian evolution. A protein-based organism in this primordial world, writes astrophysicist Christopher Chyba, could metabolise nutrients and reproduce, but not yet have the capability to encode its hereditary information within nucleic acids. Such an organism will go through a non-Darwinian evolution. If we came across something like this elsewhere (or if we could build a time machine and go back a few billion years on Earth), we would not call this life as a matter of definition.

In a different vein, offspring of modern animals that cannot themselves reproduce, such as mules, cannot undergo Darwinian evolution either and would not be 'defined' as life by NASA, but no one would argue that they are not living creatures.

Even so, says chemist Steven Benner, the definition does get around some of the confusion between 'life' and the idea of 'being alive'. 'Asking if an entity can move, eat, metabolise or reproduce might ask whether it is alive. But an individual male rabbit (for example) that is alive cannot (alone) support Darwinian evolution, and therefore is not "life".

In addition to the NASA definition, Bains adds that life forms should also contain an internal programme (on Earth, that is contained as information on DNA) that directs how the basic chemistry operates.

Together these definitions help get around the 'is fire alive?' confusion in the argument over life. According to the basic NASA definition, fire does not evolve. It might be a self-sustaining physical and chemical process but there is no internal control over how it operates, over what it consumes or how it grows. A fire tends to burn whatever combustible material is lying around in its environment. 'By contrast, the nature of a living system is determined by the coded pattern contained within that system, and is constant in a variety of environments,' says Bains. 'The living system converts energy and a variety of materials into more living systems. Growth and reproduction are chemical, but not chemically defined. There is a code, programme or blueprint that describes what each species should be like.'

Evolution, according to this idea, is not actually a requirement to make life but a consequence of the genetic code's inability to copy itself 100 per cent accurately. 'Hence it can be changed,' writes Bains. 'The genetic code is not a law of nature, but a true "program" that is interpreted by the organism. The implementation of the plan is adaptable, so that propagation can continue under changed circumstances.'

Another factor of life is that it is in disequilibrium with its environment, from a thermodynamic perspective. Thermodynamics is the study of heat and its movement. Developed to understand how to make steam engines more efficient at the time of the Industrial Revolution, it subsequently became a window into some fundamental ideas about how energy flows around the universe. Its first law showed that energy flows from hot places to cold, and its second law has become the slogan for pessimists everywhere – that the entropy of a closed system can never decrease; in other words the universe is slowly descending to chaos.

When objects are in thermodynamic equilibrium, it means that no energy flows between them – think of two objects at the same temperature. This is, of course, useless if you want to keep a living thing living. Life forms need energy to function and the complex animals and plants we see today gets their energy from food or directly from sunshine. Early life forms would have had to use simpler mechanisms, somehow exploiting natural energy imbalances (or disequilibria) in their immediate vicinity. That disequilibrium could be as simple as a difference in temperature between two places that allowed the early life form to extract energy and use it to grow and reproduce.

'Wherever a thermodynamic disequilibrium exists on Earth, organisms appear to have adapted a different metabolism to exploit it,' writes Benner. 'But these organisms all require the disequilibrium.' Understanding the thermodynamics of life is therefore useful in our search for organisms elsewhere because we can extend our list of possible search locations. Disequilibrium is common throughout the universe: the environments around stars or black holes are not in equilibrium, for example. The same is true on the Earth (and other Earth-like planets), where the nuclei of heavy elements sitting in the crust and mantle will undergo radioactive decay to release energy. That energy can drive tectonic activity and volcanoes and create non-equilibrium environments in areas such as 'black smokers' on the ocean floor, where hot water and minerals burst out of the crust and provide for life without any input from the Sun.

This thermodynamic argument for life is so powerful that Benner prefers to strip the definition of what life is and how we look for it back to something even more fundamental than the NASA definition, to something based on physical needs alone. 'The only absolute requirements', he writes, 'are a thermodynamic disequilibrium and temperatures consistent with chemical bonding.'

Benner's thermodynamic disequilibrium might be necessary for life, but life would almost certainly have needed more than that to get going. In addition to a tentative definition, therefore, we need to think about some other basic starting conditions. For a start, any molecules that are interacting to create life forms should be partly isolated from their environment so that they replicate themselves rather than getting diluted in the wider world around them (this is the basic idea behind a cell). And inside that cell there should be a medium that allows all the necessary chemical building blocks to move around each other easily and come close together in the right places and react where needed. That medium should also be able to move nutrients and information around more complex structures as the life evolves – into a body containing many cells – to keep it living.

And all of that needs to happen in a timely, reversible way, using only energies that are easily available from its environment. The medium of life should be stable and help buffer the delicate molecules of life from thermal damage from its environment – any random spikes in the temperature, up or down.

The clearest choice to do all of that is a liquid. Gas and solid media are not impossible, but the inherent difficulties are huge. Gases would not be able to hold a life form together and, though material can diffuse through solids, it might take cosmic timescales for a solid-based life form to carry out just a few metabolic reactions every few thousand years.

So far, our conditions and definitions for life have remained agnostic to the specific materials and systems needed for organising those materials. We know that life on

Earth uses water but there are plenty of liquids that could fulfil one or more of our required functions, including ammonia, methane or droplets of sulphuric acid. If these seem like odd choices right now, then that is probably a fair reaction. They seem corrosive, alien and dangerous. But to paraphrase Sagan again, that is a very hydrocentric view coming from a water-based life form.

Our impression of water as a quotidian, placid, background material is at odds with its true chemical nature. We have seen that this is an anomalous chemical but also that its strange behaviour is key to the life we experience. Step outside our profoundly grateful view of its properties, though, and it is no exaggeration to say that water is a thoroughly corrosive, reactive material. Life requires complex molecules to form and function but, as well as its myriad anomalies that make it useful for these functions, water often has an annoying tendency to cleave the chemical bonds in polymers (such as proteins), for example, as well as promoting their formation.

While most proteins cannot fold into their three-dimensional shapes without being dissolved in water, this is not necessarily a critical limitation. Molecules called peptoids, which are similar in structure to more familiar peptides in living cells, can be made to fold into their compact shapes in methanol. Biologists have also found enzymes that can work in the gas phase, and tested others successfully in non-aqueous solutions.

All of this background is useful to keep in mind when trying to choose the liquid medium for life. It is worth formulating our question as generally as possible and not just for life as we know it.

A view preferred by Andrew Pohorille, a physicist at NASA's Ames Research Center in Moffett Field, California, but often overlooked, is that living systems are distinguished by their ability to self-organise into complex structures, at many different levels, and this is achieved by associations between molecules that are not full-on chemical bonds. We've come across Pohorille's idea already in chapter five, that life is distinguished from non-life by its use of 'non-covalent interactions'. Not as strong or permanent as covalent or ionic bonds that turn collections of atoms into compounds, these interactions are key to ensuring that the functions of life can happen efficiently and without too much energy. Recall that a membrane that forms a cell wall is an assembly of molecules, such as phospholipids, kept together by non-covalent interactions. It is the same for the two helices of DNA, which are not chemically bound, though they remain close together and the interactions between them are tight and well-defined. Proteins fold into three-dimensional structures that define their functions using non-covalent interactions and enzymes in cells bring together the chemicals they work with through the same mechanism. All of these molecules store, pass on and manipulate critical information in cells and the non-covalent interactions between them are crucial to ensuring proper function.

'The essence of complex systems like living systems are those non-covalent interactions. These interactions are mediated by the solvent, to a large extent,' says Pohorille. 'Their strength would be quite different depending on the environment.'

Water is especially useful here because of its ability to form relatively weak hydrogen bonds, which are at the root of its hydrophobic effect, the most important of its suite of non-covalent interactions. 'The bottom line of this story is that the hydrophobic effect in water provides the means for matter to spontaneously organise,' says Pohorille. 'By doing so, it is this basic organising effect of life.'

The hydrogen bond is very weak in the conditions in which we find ourselves on Earth, where liquid water exists between 0 and 100°C (273 and 373K). The strong backbone of life-related bonding, when you want to build structures that will last, is the carbon–carbon covalent bond. This forms the structure of everything that makes us and these bonds are twenty times stronger than the fleeting hydrogen bonds between water molecules in a liquid. Nevertheless, as we have already seen, this weak non-covalent interaction is useful enough to give water its high boiling point, large range of temperatures over which it is a liquid and its ability to expand upon freezing into ice.

'Most of the universe does not lie between 273 and 373K, however, the range where covalent bonds are necessary and hydrogen bonds are useful,' wrote chemist Steven Benner in a paper in 2004 examining the common conditions needed for life, wherever it might be found. 'At lower temperatures, biochemistry may well be dominated by non-covalent bonds. The bonding that supports the information transfer needed for Darwinian evolution might universally mean bonding sufficiently strong at the ambient temperature to be stable for some appropriate time. In water at 273–373K, a combination of covalent bonding, hydrogen bonding and the hydrophobic effect meets this requirement. In different solvents or temperatures, different bonding may be better suited.'

Outside the Earthly conditions, other solvents can show something equivalent to the hydrophobic effect, something Pohorille generically calls the 'solvophobic' effect.

There are several possibilities that come close to mimicking the required functions of water. Liquid ammonia has similarities to water – it is a polar solvent that can dissolve organic molecules that have electrically charged regions, for example, and it can form hydrogen bonds (albeit weaker than those of water). It's not suitable for the Earth's environment because it is only liquid from -78°C to -33°C, but there are many places far out in the solar system, such as the atmosphere of Jupiter, where it is cold enough and the pressures are high enough for it to stay in a liquid phase.

At very high atmospheric pressures, around sixty times the pressure of Earth, it can remain liquid up to 100°C. Earth life could not survive in liquid ammonia, of course, and any life that does use ammonia would employ a different chemistry. Whereas life on Earth relies largely on biomolecules containing carbon-oxygen units, the equivalent for ammonia-life would rely on carbon-nitrogen units.

Another interesting feature of ammonia is that it acts as an antifreeze for water. On Titan, the surface is periodically resurfaced with a liquid that seems to have a viscosity equal to a mixture of water and ammonia. This mixture is a 'eutectic', which is something where the combination has a lower freezing point than either ingredient individually. Eutectic mixtures are common through the cosmos and can be liquid over a wide range of temperatures, making them useful for biochemistry.

Another polar solvent that has (albeit limited) potential is sulphuric acid. This is known to exist in vast amounts in the cloud layers 40–70km above the surface of Venus, making up between 80–98 per cent of the volume at different places in the form of tiny aerosol droplets. The surface temperature of Venus, around 450°C, would suggest that life would be desiccated and incapable of survival but the environmental conditions of the clouds at 50km altitude (around 37°C and 1.5 Earth atmospheres) mean that the carbon–carbon bonds familiar to Earth life could be stable. There would be plenty of energy here in the form of incoming ultraviolet radiation and a sulphuric acid medium could, like on Earth, make use of the carbon–oxygen chemical unit to build its metabolism. All of this is a stretch, to be clear, but even astronomer Carl Sagan once suggested that organisms on Venus could float above the baking surface using hydrogen bladders to keep them aloft, a bit like marine creatures on Earth use float bladders to pull themselves up and down in the seas.

Another option is formamide – chemical formula $HCONH_2$ and also known as methanamide – which is made by the reaction of hydrogen cyanide with water, both abundant in the galaxy. Like water, formamide has a big difference in electrical charge across its length – giving it the ability to form hydrogen-bonded structures with itself and other molecules in its liquid phase. It can dissolve almost anything that can dissolve in water, in particular RNA, DNA and proteins. It is not as reactive as water and, in fact, many biomolecules that are unstable in water will persist in formamide.

Suspected to exist in underground regions of Mars, formamide is a liquid over a wide range of environmental conditions; it can dissolve ionic salts such as sodium chloride and shows some solvophobic effects. It is quickly broken up in the presence of water, however, and would only be expected in dry, desert-like environments.

We have already seen, in McKay's proposed film-like life form, that oceans of liquid hydrocarbons such as methane and ethane could be the solvent of life on Titan. Unlike the previous examples, these hydrocarbons are non-polar, which means they will not be as good as water at dissolving the ionic salts that are used in Earth-based life for many specific functions including signalling between cells and the transfer of electrical energy. But who is to say that life on Titan needs this mechanism of chemicals to live anyway?

Hydrocarbons are abundant throughout the solar system and there are several places where the temperatures and pressures are right for oceans to exist. Broadly speaking, these chemicals could be as versatile as water in enabling the reactions between organic molecules required by life forms, says Benner, and they could display something similar to a solvophobic effect that would allow the hydrocarbon building blocks of living things to self-organise to some degree. In this case, hydrocarbons with polar sections along their length could demonstrate a 'hydrocarbon-phobic' effect. Pohorille is less sure hydrocarbons alone would be any use in this context, adding that they are usually not very good solvents without the presence of water. But perhaps hydrocarbon-water mixtures might work for life.

But we already know how good hydrocarbon solvents can be for running reactions between organic molecules because chemists commonly use them as a medium for their experiments in the lab. Chemists often avoid water as a medium because it is so reactive, tearing many organic molecules apart even when it is meant to be sitting in the background. All of this adds to the case that Titan is fit for life, given its appropriate environmental conditions, availability of solvent and source of raw ingredients (such as metallic rocks and surfaces where life could start).

If McKay's sheet-like organism seems like a fantasy too far, it is worth noting that hydrocarbons might also provide a useful background to water-based life. We would expect water to be frozen if it appeared on the surface of somewhere like Titan but some astronomers have argued that liquid water might exist there, heated by the after-effects of impacts from comets and asteroids. If this is the case, life would have the same opportunities and constraints it does anywhere else there is water. In addition, water mixed with a hydrocarbon solvent on Titan would form tiny droplets (like oil forms droplets if mixed with water on Earth) and these bubbles could provide the necessary isolation for Darwinian evolution, an alternative to the fatty cell membranes we see on Earth that provide barriers between the inside and outsides of a biological cell (in our case, a layer of biomolecules separating two areas of water).

There are even more exotic options such as liquid nitrogen or bizarre, supercritical liquid states of hydrogen and helium that might occur in the clouds of the giant gas planets in the outer solar system. Liquid hydrogen is an attractive idea because it is by far the most common molecule in the solar system and more than 85 per cent of the atmospheres of Jupiter, Saturn, Uranus and Neptune are made from hydrogen molecules. Around 14 per cent of the outer parts of these planets is helium, with smaller amounts of methane, water, ammonia and hydrogen sulphide. Above particular temperatures and at extremely high pressures in the atmospheres of these gas giants, hydrogen and helium enter a supercritical phase that is neither liquid nor gas. This phase is known to have strange properties and, though a neat idea as a medium for life, very little is understood about how organic molecules might behave in supercritical fluids.

'Throughout most of the volume of gas giant planets where molecular dihydrogen is stable, it is a supercritical fluid,' writes Benner. 'For most of the volume, however, the temperature is too high for stable carbon–carbon covalent bonding. We may, however, define two radii for each of the gas giants. The first is the radius where dihydrogen becomes supercritical. The second is where the temperature rises to a point where organic molecules are no longer stable; for this discussion, this is chosen to be 500K. If the second radius is smaller than the first, then the gas giant has a "habitable zone" for life in supercritical dihydrogen. If the second radius is larger than the first, however, then the planet has no habitable zone.'

Any such habitable zone on Jupiter would be small – in places where organic molecules could survive, around 300K, the atmospheric pressure would be around 8 Earth atmospheres, which is too low for hydrogen to become supercritical. 'For Saturn, Uranus and Neptune, however, the habitable zone appears to be thicker (relative to the planetary radius),' writes Benner. 'On Saturn, the temperature is circa 300K when dihydrogen becomes supercritical. On Uranus and Neptune, the temperature when dihydrogen becomes supercritical is only 160K; organic molecules are stable at this temperature.'

Another issue is that the atmospheres of the gas giants convect. In order to survive on Jupiter, Benner says, any hypothetical life would have to avoid being moved around by the air currents into hostile places. This is also true for anything living in fluids, of course; even the Earth's oceans and organisms evolved in these places might have found ingenious solutions.

It's one thing to theoretically imagine a solvent or mixture doing some of the chemical functions of water but another important factor is whether your solvophobic-capable liquid occurs in large enough quantities in space. One of water's greatest advantages in the formation of life is, put simply, the fact that there is so much of it in the universe. If the chosen alternative is rare, then it can only ever support life in a theoretical sense, or we need to find exoplanets that can support its presence as a potential solvent – decades away at least.

Hydrogen is by far the most common element in the universe, followed by helium, oxygen, neon, carbon, nitrogen, silicon, argon and phosphorus. That availability of elements drops off quickly – relative to hydrogen, there is around 14 per cent as much oxygen, around 10 per cent as much neon, 8.7 per cent as much carbon, 5.6 per cent as much nitrogen and only 0.07 per cent as much phosphorus. Crudely speaking, we can expect that the frequency of stable compounds between these elements would follow a similar order – H_2O will be far more common than NH_3 (ammonia), and both will be much more common than other compounds of nitrogen and oxygen. (This relative abundance is indeed seen in interstellar ices and also in near-Earth comets.)

Another way to test some of these ideas is to use the leading edge of research into designed life, synthetic biology, a discipline where one of the explicit aims is to build a new life form from scratch, starting from its chemical parts. Efforts so far have tried to build cell-like structures and to modify the known chemicals of life – water, DNA and so on. There is work going on with alternative chemicals within the proteins and the bases that encode information in DNA. These techniques are ripe for constructing alternative chemistries using solvents other than water and biomolecules other than the proteins, sugars and lipids that we know.

There may also be limitations beyond the control of the specific chemical reactions that are possible with a solvent. Life began on Earth relatively soon after the hellish environment and oceans settled, less than a billion years after the formation of our planet. This was not the complex life we know today, which took many more billions of years of evolution, but the earliest living systems that could grow and replicate. They would have not had the chemical catalysts modern life uses, enzymes, because they had not yet evolved to that level.

Instead, they would have used heat to speed up their nascent biochemical reactions by driving up the chemical kinetics to a level where they could sustain a population, rather than dying off or becoming diluted too quickly. According to Richard Wolfenden at the University of North Carolina, biochemical reactions in this pre-complex-life world would have been speeded up, for example, 10 million-fold with a temperature change from 25°C to 100°C. This is consistent with what is thought to be the leading candidate for how life started on Earth, at the bottom of the oceans next to boiling-hot hydrothermal vents where there are copious supplies of nutrients and energy. This pseudo-catalytic help from heat would have reduced the time needed to set up a primitive biochemistry from millions to tens of years.

If that scenario about the first moments of life is correct, then it severely limits the solvents that can only remain liquid at frigid temperatures, where the chemical kinetics are necessarily slower. In this scenario, if kinetics is important for the first flicker of life, cold oceans of liquid ammonia, formamide or hydrocarbons might be possible as cradles for life but they might just be too slow for the chemistry to turn into biochemistry.

The US National Research Council panel that considered alternatives to water concluded that life should be possible in forms very different from that which we see on Earth. 'Different specific biomolecules may be considered highly likely in extra-terrestrial life. Different architectures at the microscopic and macroscopic levels must also be considered likely,' they wrote. 'Furthermore, life should be considered possible in aqueous environments that are extreme in their solute content, in their acidity or alkalinity, and in their temperature range, especially with ammonia as an antifreeze in low-temperature water-ammonia eutectics. [We see] no reason to exclude the possibility of life in environments as diverse as the aerosols above Venus and the water-ammonia eutectics of Titan.'

All of this is the source of much debate, of course. Pohorille recalls sitting on a panel at a scientific conference with Benner, who has spent decades in his research career examining and promoting alternatives to water in possible chemistries for life on other worlds. After a rousing presentation from Benner about how unsuitable water theoretically should be for life, given its corrosiveness and ability to tear apart certain organic molecules, Pohorille was momentarily convinced of the anti-water position. 'Benner made a great case that water is, from a perspective of an organic chemist, not a very good solvent for life,' he says. 'He was so convincing that I came to the conclusion that he is indeed right and that life could not have arisen in water and therefore we don't exist. It's just an illusion that we are here and talking.'

Pohorille is only partly joking. We have come to accept

water as the medium of life for sensible reasons, many of which we have traced in this book. But if there are any lessons in the human story of water – its anomalous nature, its key role in the evolution and functions of life and civilisation and its ubiquity in the universe – it is surely that, in our privileged position as the products of water, we should be cautious in assuming that anything, water's singular role included, is as simple or definite as it might initially seem.

Epilogue

It will come as no surprise that our Christmas in Antarctica was white. The pack ice around our ship, the *Akademik Shokalskiy*, stretched in an unending grey-white field to the horizon on all sides that morning, a flat expanse of frozen water covered in boulders of ice with, inching along in the distance here and there, icebergs. Icicles hung from the metal steps of the ship's exterior and the decks were covered in flurries of snow. Two weeks after setting sail from Port Bluff in New Zealand, we had arrived at Cape de la Motte near the Mertz glacier (named after Douglas Mawson's trekking colleague who had succumbed to the ravages of the continent) a day earlier and we had planned to stay for a while to explore the nearby Hodgeman Islands before hauling anchor and heading home, back to the comforts of civilisation. But nothing goes quite as you would want on this mighty continent.

On that festive morning, we were beset with ice. The captain had tried hard to force the ship into open water overnight but the ice had been too strong, too thick to let us through. The south-westerly winds had pushed the ice floes against the continent and the *Shokalskiy* had been pinned in place, as if in a vice. We were stuck, as ships often are when ice conditions change quickly on the Antarctic coast, forced into a pause on our expedition as we waited for the polar winds to shift and blow a hole in the frozen sea around us. The continent had us in its grasp and, though help had been summoned and was coming in the form of several icebreakers that promised to get us out, the continent would decide when to let us go.

While we waited in the 50-knot winds, the science went on. Oceanographers used the hiatus to drill holes through the ice and drop temperature and salinity probes into the frigid ocean below. The marine ecologists carried on recording the sounds underwater to work out what was living there and the ornithologists continued their hourly watch from the ship's bridge.

In the following days, our situation got worse. When we had first become beset, we could see the open ocean beyond the ice field, a mere 3km away. A few days later, the winds had whipped even more ice into place and the nearest open water was more than 30km away. In addition, several icebergs had started to move near the ship. While pack ice is moved around by the wind, bergs move instead with the ocean currents because they stretch for hundreds of metres under the water's surface. That means these behemoths can cut through fields of ice as if they were open ocean. While they could move through the frozen sea as if it weren't there, our ship could not. Our captain called for help as soon as the icebergs hove into view. If one of them had come towards us, our relatively tiny vessel would have stood no chance in its path.

The rescue wasn't easy. A powerful Chinese icebreaker, the *Xue Long*, came within six nautical miles of our ship and the mood aboard the *Shokalskiy* lifted on the evening when we saw the ship as a faint red dot on the horizon. Unfortunately it could not break through the thick ice overnight and had to turn back. A French vessel, *L'Astrolabe*, had already arrived at the sea ice edge at the same time as the *Xue Long* but had avoided going into the ice floes, since its icebreaking capability was less than that of the Chinese vessel. Our hopes then had to rest on the Australian icebreaker, the *Aurora Australis*, which was still several days away at sea.

Ten days after we got stuck, the captains of the three ships decided to evacuate us by helicopter. In a tense, six-hour operation, groups of twelve of us trudged across an ice floe next to our ship that had been marked out as a makeshift helipad and where the *Xue Long*'s twin-blade helicopter set itself down. On a clear Antarctic evening, we flew 35km to the *Aurora Australis*, which was moored up against the edge of the vast field of ice. The ice had been so thick, so extensive that the Australian vessel had not even tried to go in, fearing it might get trapped – the *Xue Long* was already having trouble manoeuvring out from its position after entering the ice field.

From the air, even the Antarctic veterans gasped at the extent of the ice that had closed in around us in such a short amount of time. Our blue ship was just a speck on the vista of pure white, invisible within minutes of us getting in the air.

What caused the sudden build-up of ice is not clear. The ice that surrounded the *Shokalskiy* was not the fresh ice you see forming every winter, brittle and a metre or so thick. Instead it looked as though it had frozen several winters ago and hardened and thickened over several more years. The most likely hypothesis among the scientists on board our ship was that it had formed elsewhere along the coast, further south where it was even colder, and then been broken off in a recent heavy storm. Those pieces of multi-year ice had then been blown to Cape de la Motte, just as we had been intending to leave, and pinned us in place.

After sailing through and then being embedded in it, looking at the ice from the air gave each of us a qualitatively different (and each of us unique) new experience of the ice. For twenty minutes on the helicopter, we had clear views in every direction for dozens of miles. This relatively tiny chunk of the cryosphere still vastly outmatched our puny human instruments and machines and, yet, this was the part of the world that was most affected by the life we all led elsewhere in the warmer, more comfortable parts of the world. The recent adventure had shown how our efforts, dreams, development and mastery of the Earth might not work all the time but, inexorably, those collective efforts were chipping away at this very landscape in measurable, irreversible ways.

The vast ice fields of Antarctica are a salutary reminder that this is the form water takes, for the most part, in the universe. Frozen, inhospitable, barren, alien. Our relatively insignificant pool of liquid water on Earth (and perhaps others in places we have yet to find) is a rare outlier of a substance that exists as solid crystals virtually everywhere that it occurs. We humans have taken that liquid to mean everything, of course. It has been our wellspring of stories and culture, the source of creation and death; it has shaped our language and politics and been at the core of how we built civilisation.

All of the effects we have on the water on our planet will be ephemeral. These molecules were there to help create the Earth and the life on it, and we have built our worlds around it. We should rightly worry about how we are changing the natural balance of its forms on our planet because the ratios of solid, liquid and gaseous water fundamentally affects our lives and those of everything else that lives now and that will live in the future. But the form of the water – whether in gigantic ice sheets on Antarctica, lapping against the shores of our cities or constrained within biological cells – is just a momentary expression of changing energy in this remarkable chemical's path through the universe. We might be inexorably tied to water and its forms, but water is not tied to us.

The ingredients of water in us and around us existed in the first few million years after the Big Bang and they will persist long after they have made their way through the story I have traced in this book. The universe has countless billions of years ahead of it, aeons in which water will return to space long after humans have gone, after the Earth has gone, after the Sun is ashes.

In around 1 billion years' time, a quarter of the world's oceans will have been absorbed into the mantle. The Sun's luminosity will steadily increase and, by the time it is 10 per cent brighter than it is today, the surface of the Earth will be an average of 47°C. The greenhouse conditions will make life unbearable for the most part and the highest parts of the atmosphere, the stratosphere, will contain ever more water. Sunlight hitting this part of the atmosphere will split water atoms, allowing the free hydrogen to escape the Earth. Over the following 100,000 years, the world's seas will disappear. Water will still sit in pools on the surface as it is released from inside the Earth, and there might be some small lakes at the poles. Eventually, by the time the Sun has run out of hydrogen fuel in 5 billion years, any remaining water on the Earth's surface will have been driven away by the extreme temperatures. Disassociated into its constituent elements, all the water we have ever known will be atoms, floating in the blackness of space.

Acknowledgements

Where I have discussed the results of specific research projects, papers or conversations, I've usually credited the appropriate people in the main text or in the notes. But, in addition, the following people gave me a lot of their time to help me grapple with some of the more technical and structural aspects of the book: Andrea Sella, Angelos Michaelides, Christoph Salzmann, Roger Masters, David Bodanis, Ewine van Dishoeck, Peter Gleick, Hasok Chang and Philip Ball.

Thank you also to the scientists and press officers at NASA's Ames and Jet Propulsion Laboratories in California, who let me into their world for several days during the summer of 2014 to show me how they send robots onto other worlds and why they do it. In particular, my many conversations and meetings with Chris McKay, Bonnie Buratti, Lynn Rothschild, Michael Russell and Andrew Pohorille did a lot to help shape the final sections of the book.

Robbert Dijkgraaf at the Institute for Advanced Study in Princeton invited me to spend several weeks at Fuld Hall in the summer of 2014 and Christine Ferrara in the institute's media office made my initial introductions with him. Thank you both for giving me a space in which to focus on the writing and for lending me an office that was just five doors down from the late Albert Einstein's.

The Antarctic sections of this book were written on the Australasian Antarctic Expedition, which set sail from New Zealand in December

2013. Thank you to expedition leader Chris Turney for inviting me onboard and co-leader Greg Mortimer for leading us out of trouble after we were beset. Ben Maddison, Andrew Peacock and the passengers and crew of the *Academic Shokalskiy* were the finest people with which you could get stuck in a remote location – thank you all for your friendship and stories. And thank you to the crews of the icebreakers *Xue Long, L'Astrolabe* and *Aurora Australis* for making such great efforts to rescue us, at great time cost to yourselves.

My editors at the *Guardian*, Alan Rusbridger and Emily Wilson, have provided me with guidance and encouragement for more than a decade. Thank you both for giving me that first chance all those years ago. They and my colleagues Ian Sample, James Randerson, Laurence Topham, James Kingsland, Mustafa Khalili and Sarah Hewitt also have my thanks for their support and comradeship over so many years. I'd also like to thank Geoff Hill and Tim Singleton at ITV News, who allowed me start my new job with them several months late, so that I had the time to finish this book after leaving the *Guardian*.

Simon Thorogood, my editor at Headline, has been at every twist and turn in the creation of this book – thank you for guiding me through this adventure.

This book would never have happened without the energy and commitment of my literary agent, Will Francis. Over a lunch on High Street Kensington several years ago, we discussed a few vague ideas for a book about water and he persuaded and nudged me to turn it into reality.

Thanks also to everyone who helped copyedit, design and prepare this book for publication. Any errors remaining in the text are mine alone.

My biggest thanks go to those who supported and encouraged me during the long, long process of taking a book from idea to the object you hold in your hands: my parents, brother, friends and, in particular, Cat de Lange for her patience, for being the first reader and for helping me make it far better than I could have done on my own.

Alok Jha, January 2015

Notes and References

For each chapter, I have included specific references for the text and, after that, an alphabetical list of sources that were useful for general background. Many sources were useful for more than one chapter and, in that case, I've listed full details where they were first used and used abbreviated references in later chapters.

Introduction

'One of the roots for water . . .':
Tulloch, Alexander (2008), 'Water, Water Everywhere . . .', *English Today* 24, pp. 62–4, DOI: 10.1017/S026607840800031X

PART I: HYDROSPHERE

Chapter One: Departures

'Through aeons of cycles of freezing and melting':
Franks, Felix (1972), 'Introduction – Water, The Unique Chemical', from *Water, a Comprehensive Treatise: The Physics and Physical Chemistry of Water, Volume 1*, Plenum Press (New York)

'All of the water on our planet should exist as vapour':
Ball, Philip (2002), 'Water – Life's Matrix: Extended', http://www.philip-ball.co.uk/index.php?option=com_content&view=article&id=56:ater-lifes-matrix-extended&catid=13:water&Itemid=16

'The Mpemba Effect':
Mpemba, Erasto and Osborne, Denis (1969), 'Cool?', *Physics Education* 4, 172, DOI: 10.1088/0031-9120/4/3/312

Jeng, Monwhea (2006), 'The Mpemba Effect: When Can Hot Water Freeze Faster than Cold?', *American Journal of Physics* 74, 514, DOI: 10.1119/1.2186331

'These anomalous behaviours of water are due to the intriguing ability of its molecules . . .':
Ball, Philip (2011), 'Why Water Is Weird', a talk delivered at the Royal Society of Chemistry in London, http://www.philipball.co.uk/images/stories/docs/pdf/RSC_water_talk+pics_2.pdf

General sources:
Ball, Philip (2004), 'Water, Water, Everywhere?', *Nature* 427, pp. 19–20, DOI: 10.1038/427019a

Buswell, Arthur and Rodebush, Worth (1956), 'Water', *Scientific American* 194, pp. 76–92; DOI: 10.1038/scientificamerican0456-76

Finney, John (2004), 'Water? What's So Special About It?', *Philosophical Transactions of the Royal Society of London B* 359, pp. 1145–65; DOI: 10.1098/rstb.2004.1495

Huang, C. et al. (2009), 'The Inhomogeneous Structure of Water at Ambient Conditions', *Proceedings of the National Academy of Sciences* 106, pp. 15214–8, DOI: 10.1073/pnas.0904743106

Chapter Two: Southern Ocean

For more on the science goals of the six-week Australian Antarctic Expedition 2013–14, led by climate scientist Professor Chris Turney of the University of New South Wales, go to 'The Science Case' at http://www.spiritofmawson.com/the-science-case/

The Spirit of Mawson website also has more details of the route of the expedition, including points of interest. I wrote a detailed daily diary of the expedition for *Guardian* here: http://www.theguardian.com/science/antarctica-live

Chapter Three: On the Origin of Water

'The Egyptians told of the time before creation . . .':
There are many books and articles that tell ancient origin stories, not least the sacred texts of the world's religions. For this section, I was helped by conversations with Professor Michael Witzel of Harvard University, who also kindly shared his essay 'Water in Mythology' before its publication.

'Using the Herschel space observatory, scientists have seen these stellar firehoses in action':
Kristensen, Lars et al, 'Water in Low-mass Star-forming Regions with Herschel (WISH-LM), High-velocity H_2O Bullets in L1448-MM Observed with HIFI', *Astronomy & Astrophysics* 531, L1, DOI: 10.1051/0004-6361/201116975

'In this nascent solar system, beyond the Sun, individual molecules and atoms floated . . .':
This story of the creation of water in space was based on interviews, articles and research papers on the work of astronomer and chemist Professor Ewine van Dishoeck and her team at the Leiden Observatory in the Netherlands.

Van Dishoeck, Ewine et al. (2014), 'Water: From Clouds to Planets', accepted for publication in *Protostars and Planets VI*, http://arxiv.org/abs/1401.8103

Van Dishoeck, Ewine et al. (2011), 'Water in Star-forming Regions with the Herschel Space Observatory (WISH). I. Overview of Key Program and First Results', *Publications of the Astronomical Society of the Pacific* 123, pp. 138–170, DOI: 10.1086/658676

Van Dishoeck, Ewine (2011), 'Water in Space', *Europhysics News* 42, pp. 26–31, DOI: 10.1051/epn/2011105

Bergin, Edwin and van Dishoeck, Ewine (2012), 'Water in Star- and Planet-forming Regions', *Philosophical Transactions of the Royal Society A* 370, pp. 2778–2802, DOI: 10.1098/rsta.2011.0301

'We do have a restricted window into this time' and 'No one knows how many objects hit the Earth and how much water':
Kasting, James (2003), 'The Origins of Water on Earth', *Scientific American* 13, 28–33, DOI: 10.1038/scientificamerican0903-28sp

Kasting, James, Toon, Owen & Pollack, James (1998), 'How Climate Evolved on the Terrestrial Planets', *Scientific American* 258, 90–97, DOI: 10.1038/scientificamerican0288-90

'A wrinkle in the comet theory . . .':
Altwegg, Kathrin et al. (2014), '67P/Churyumov-Gerasimenko, a Jupiter Family Comet with a High D/H Ratio', *Science Express*, DOI: 10.1126/science.1261952

General sources:
Akeson, Rachel (2011), 'Watery Disks', Science 334, pp. 316–7, DOI: 10.1126/science.1213752

Allègre, Claude and Schneider, Stephen (1994), 'The Evolution of the Earth', *Scientific American* 271, pp. 66–75, DOI: 10.1038/scientific american1094-66

Blake, David and Jenniskens, Peter (2001), 'The Ice of Life', *Scientific American* 285, pp. 44–51, DOI: 10.1038/scientificamerican0801-44

Drake, Michael (2005), 'Origin of Water in the Terrestrial Planets', *Meteoritics & Planetary Science* 40, pp. 519–27, DOI: 10.1111/j.1945-5100.2005.tb00960.x

Hoffman, Paul and Schrag, Daniel (2000), 'Snowball Earth', *Scientific American* 282, pp. 68–75, DOI: 10.1038/scientificamerican0100-68

Lowenstein, T. et al. (2014), 'The Geologic History of Seawater', *Treatise on Geochemistry (Second Edition)*, pp. 569–622, DOI: 10.1016/B978-0-08-095975-7.00621-5

Mottl, Michael et al. (2007), 'Water and Astrobiology', *Chemie der Erde – Geochemistry* 67, pp. 253–82, DOI: 10.1016/j.chemer.2007.09.002

Robert, François (2001), 'The Origin of Water on Earth', Science 293, pp. 1056–8, DOI: 10.1126/science.1064051

Sleep, N. H. (2001), 'Initiation of Clement Surface Conditions on the Earliest Earth', Proceedings of the National Academy of Sciences 98, pp. 3666–72, DOI: 10.1073/pnas.071045698

Chapter Four: The Hydrosphere

On the Challenger Expedition:
'History of Oceanography: *HMS Challenger*', http://aquarium.ucsd.edu/Education/Learning_Resources/Challenger/introduction.php

'The History of Ocean Exploration', Natural History Museum website, http://www.nhm.ac.uk/nature-online/science-of-natural-history/expeditions-collecting/hms-challenger-expedition/exploration-history/index.html

'History of the Challenger Expedition', Challenger Society for Marine Science website, http://www.challenger-society.org.uk/about-us/history-challenger-expedition

'Around 71 per cent of the Earth's surface is covered in water, to an average depth of 3,700m':
Pörtner, Hans and Karl, David et al. (2014), 'Ocean Systems', Chapter 6, from the report prepared by Working Group II for the Fifth Assessment Report of the Intergovernmental Panel on Climate Change, http://ipcc-wg2.gov/AR5/images/uploads/WGIIAR5-Chap6_FGDall.pdf

'If all the world's ice and snow were melted and spread evenly . . .:
Penman, H. L. (1970), 'The Water Cycle', *Scientific American* 223, pp. 98–108, DOI: 10.1038/scientificamerican0970-98

The hydrological cycle:
Ball, Philip (2000), 'The Hydrological Cycle', *Nature*, DOI: 10.1038/news000127-12

Chahine, Moustafa (1992), 'The Hydrological Cycle and Its Influence on Climate', *Nature* 359, pp. 373–80, DOI: 10.1038/359373a0

Dooge, James (2001), 'Concepts of the Hydrological Cycle', remarks delivered at the International Symposium H_2O 'Origins and History of Hydrology' in Dijon

Hubbart, J. (2008), 'History of Hydrology', http://www.eoearth.org/view/article/153525

Nace, Raymond (1975), 'The Hydrological Cycle: The Historical Evolution of the Concept', *Water International* 1, pp. 15–21, DOI: 10.1080/02508067508685694

The Gulf Stream:
'1785: Benjamin Franklin's "Sundry Maritime Observations"', NOAA Ocean Explorer website, http://oceanexplorer.noaa.gov/history/readings/gulf/gulf.html

'Gulf Stream', Encyclopædia Britannica Online, http://www.britannica.com/EBchecked/topic/249180/Gulf-Stream

'Great North Eastern Garbage Patch':
McMahon, Bucky (2014), 'The Terrifying True Story of the Garbage that Could Kill the Whole Human Race', *Matter*, https://medium.com/matter/the-terrifying-true-story-of-the-garbage-that-could-kill-the-whole-human-race-b17eebd6d54

'The world of deep ocean currents':
Lozier, Susan (2010), 'Deconstructing the Conveyor Belt', *Science* 328, pp. 1507–11, DOI: 10.1126/science.1189250

NOAA factsheet on tides and currents, http://oceanservice.noaa.gov/education/kits/currents/lessons/currents_tutorial.pdf

Orsi, Alejandro (2010), 'Recycling Bottom Waters', *Nature Geoscience* 3, pp. 307–9, DOI: 10.1038/ngeo854

'Thermohaline Circulation', Encyclopædia Britannica Online, http://www.britannica.com/EBchecked/topic/591633/thermohaline-circulation

Thomson, Jeremy (2000), 'Great Rivers of the Ocean', *Nature*, DOI: 10.1038/news010111-6

Stewart, R. W. (1969), 'The Atmosphere and the Ocean', *Scientific American* 221, pp. 76–86, DOI: 10.1038/scientificamerican0969-76

Wunsch, Carl (2002), 'What Is the Thermohaline Circulation?', *Science* 298, no. 5596, pp. 1179–81, DOI: 10.1126/science.1079329

'Evidence from chemicals in the fossilised shells . . .':
Peplow, Mark (2006), 'Ocean Currents Flip Out', *Nature*, DOI: 10.1038/news060102-5

'Decided ... to retrace the path of the HMS Challenger':
Roemmich, Dean et al. (2012), '135 Years of Global Ocean Warming Between the Challenger Expedition and the Argo Programme', *Nature Climate Change* 2, pp. 425–8, DOI: 10.1038/NCLIMATE1461

IPCC report:
Hoegh-Guldberg, Ove and Cai, Rongshuo et al. (2014), 'The Ocean', Chapter 30 from the report prepared by Working Group II for the Fifth Assessment Report of the Intergovernmental Panel on Climate Change (IPCC), http://ipcc-wg2.gov/AR5/images/uploads/WGIIAR5-Chap30_FGDall.pdf

Most of the data and numbers in the climate change section of this chapter are taken from Chapters 6 and 30 of the IPCC's Working Group II analysis of the Fifth Assessment Report. This represents the best available peer-reviewed evidence on the state of the marine environment and the likely impacts of a warming world.

'Coral reefs are the Earth's most diverse marine ecosystem':
Jha, Alok (2011), 'Conservationists Unveil Plans to Save Coral from

Extinction', *Guardian*, http://www.theguardian.com/environment/2011/jan/11/plans-save-coral-extinction/

'In 2008, a group of artisan miners found a tiny, battered-looking diamond':
Pearson, Graham et al. (2014), 'Hydrous Mantle Transition Zone Indicated by Ringwoodite Included Within Diamond', *Nature* 507, pp. 221–4, DOI: 10.1038/nature13080

Sample, Ian (2014), 'Rough Diamond Hints at Vast Quantities of Water Inside Earth', *Guardian*, http://www.theguardian.com/science/2014/mar/12/rough-diamond-water-earth-wet-zone

General sources:
Herring, David (1999), 'Ocean and Climate', NASA Earth Observatory website, http://earthobservatory.nasa.gov/Features/OceanClimate/oceanclimate.php

Langenberg, Heike (2012), 'Hydrology: Complex Water Future', *Nature Geoscience* 5, p. 849, DOI: 10.1038/ngeo1658

Rahmstorf, Stefan (2003), 'The Current Climate', *Nature* 421, p. 699, DOI: 10.1038/421699a

Smithsonian Institution website, 'Currents, Waves, and Tides: The Ocean in Motion', http://ocean.si.edu/ocean-news/currents-waves-and-tides-ocean-motion

Stockholm International Water Institute website, http://www.siwi.org/media/facts-and-statistics/1-water-resources-and-scarcity

Visbeck, Martin (2007), 'Oceanography: Power of Pull', *Nature* 447, p. 383, DOI: 10.1038/447383a

PART II: BIOSPHERE

Chapter Five: Life

'The phytoplankton are a multitude of species . . .':
Isaacs, John (1969), 'The Nature of Oceanic Life', *Scientific American* 221, pp. 146–62, DOI: 10.1038/scientificamerican0969-146

'Water . . . acts as a shock absorber and a building material':
Jequier, Eric et al. (2010), 'Water as an Essential Nutrient: the Physiological Basis of Hydration', *European Journal of Clinical Nutrition* 64, pp. 115–23, DOI: 10.1038/ejcn.2009.111

Rush, Elaine (2013): 'Water: Neglected, Unappreciated and Under Researched', *European Journal of Clinical Nutrition* 67, pp. 492–5, DOI: 10.1038/ejcn.2013.11

Fitzsimons, James (1976), 'The Physiological Basis of Thirst', *Kidney International* 10, pp. 3–11, DOI: 10.1038/ki.1976.74

'Water is no background player . . .':
Published interview with Philip Ball (2003), 'Water: The Molecule of Life,' *Astrobiology*, http://www.astrobio.net/interview/453/water-the-molecule-of-life#

'Across all domains of life, the amount of water contained within varies . . .':
Franks (1972) – see Chapter One

'The body can use the last of these as an emergency source of water . . .':
Wolf, A. V. (1958), 'Body Water', *Scientific American* 199, pp. 125–35, DOI: 10.1038/scientificamerican1158-125

'Molecular life began bathed in it . . .':
Rand, Peter (2004), 'Probing the Role of Water in Protein Conformation and Function', *Philosophical Transactions of the Royal Society of London B* 370, pp. 1277–85, DOI: 10.1098/rstb.2004.1504

The water controversy:
There are several books about this intriguing period of history and the
best place to start is this essay: Miller, David Philip (2002), 'Distributing
Discovery between Watt and Cavendish: A Reassessment of the
Nineteenth-Century "Water Controversy"', *Annals of Science* 59, pp.
149–78, DOI: 10.1080/00033790110044747. See also:

Jacobsen, Anja Skaar (2006), 'The Water Controversy', *Minerva* 44, pp.
459–62, DOI: 10.1007/s11024-006-9011-1

Thorpe, T. E. (1898), 'James Watt and the Discovery of the Composition
of Water', *Scientific American* 46, pp. 18900–2, http://www.nature.com/
scientificamerican/journal/v46/n1179supp/pdf/scientificamer-
ican08061898-18900supp.pdf

Cavendish, Henry (1766), 'Three Papers, Containing Experiments on
Factitious Air, by the Hon. Henry Cavendish, FRS', *Philosophical
Transactions of the Royal Society of London* 56, DOI: 10.1098/
rstl.1766.0019

Cavendish, Henry (1784), 'Experiments on Air', *Philosophical
Transactions of the Royal Society of London* 74, pp. 119–53, DOI:
10.1098/rstl.1784.0014

Watt, James (1784), 'Thoughts on the Constituent Parts of Water and of
Dephlogisticated Air; With an Account of Some Experiments on That
Subject. In a Letter from Mr. James Watt, Engineer, to Mr. De Luc, FRS',
Philosophical Transactions of the Royal Society of London 74, pp. 329–53,
DOI: 10.1098/rstl.1784.0026

Watt, James (1784), 'Sequel to the Thoughts on the Constituent Parts of
Water and Dephlogisticated Air. In a Subsequent Letter from Mr. James
Watt, Engineer, to Mr. De Luc, FRS', *Philosophical Transactions of the
Royal Society of London* 74, pp. 354–7, DOI: 10.1098/rstl.1784.0027

Muirhead, James Patrick (ed.) (1846), *Correspondence of the Late James
Watt on his Discovery of the Theory of the Composition of Water*, John
Murray (London)

Also helpful to understand the meaning of 'discovery' in the early scientific context and an examination of why it took quite so long for the thinkers of the time to finally get to the H_2O formula: Chang, Hasok (2012), 'Is Water H_2O?: Evidence, Realism and Pluralism', *Boston Studies in the Philosophy of Science* 293, pp. 1–70, DOI: 10.1007/978-94-007-3932-1_1

Chang, Hasok (2012), 'Water: The Long Road from Aristotelian Element to H_2O', *Circumscribere* 12, pp. 1–15, http://revistas.pucsp.br/index.php/circumhc/article/view/13401/9934

'By the time of the Karlsruhe Congress of 1860 . . .':
Everts, Sarah (2010), 'When Science Went International', *Chemical & Engineering News*, https://pubs.acs.org/cen/science/88/8836sci1.html

'Water's polarity makes it a fantastic solvent':
Gerstein, Mark and Levitt, Michael (1998), 'Simulating Water and the Molecules of Life,' *Scientific American* 279, pp. 100–105, DOI: 10.1038/scientificamerican1198-100

'The more molecules there are in a cluster, the stronger the overall cohesion':
Chaplin, Martin (2001), 'Water: Its Importance to Life', *Biochemistry and Molecular Biology Education* 29, pp. 54–9, DOI: 10.1016/S1470-8175(01)00017-0

Chaplin, Martin (2006), 'Do We Underestimate the Importance of Water in Cell Biology?', *Nature Reviews Molecular Cell Biology* 7, pp. 861–6, DOI: 10.1038/nrm2021

'These non-covalent interactions can actually define a living cell':
Pohorille, Andrew and Pratt, Lawrence (2012), 'Is Water the Universal Solvent for Life?', *Origins of Life and Evolution of Biospheres* 42, pp. 405–9, DOI: 10.1007/s11084-012-9301-6

The Murchison meteorite:
The Meteoritical Society – entry for the Murchison meteorite, http://www.lpi.usra.edu/meteor/metbull.php?code=16875

'Deamer took a small sample of the Murchinson rock . . .':
Deamer, David (1985), 'Boundary Structures are Formed by Organic Components of the Murchison Carbonaceous Chondrite', *Nature* 317, pp. 792–4, DOI: 10.1038/317792a0

'In the early 1960s, Bangham found that ovolecithin . . .':
Düzgüneş, Nejat and Gregoriadis, Gregory (2005), 'Introduction: The Origins of Liposomes: Alec Bangham at Babraham', *Methods in Enzymology* 391, pp. 1–3, DOI: 10.1016/S0076-6879 (05)91029-X

'Instead, Bangham and Deamer thought that if fatty molecules existed . . .':
Zimmer, Carl (1985), 'First Cell', *Discover Magazine*, http://discovermagazine.com/1995/nov/firstcell584

A few more interesting papers from Bangham and Deamer's work on cell membranes over several decades: Deamer, David et al. (2002), 'The First Cell Membranes', *Astrobiology* 2, pp. 371–38, DOI: 10.1089/153110702762470482

Bangham, A. D. and Horne, R. W. (1964), 'Negative Staining of Phospholipids and their Structural Modification by Surface-active Agents as Observed in the Electron Microscope', *Journal of Molecular Biology* 8, pp. 660–8, DOI: 10.1016/S0022-2836(64)80115-7

'Deamer took the idea further . . .':
Deamer, David (2010), 'From "Banghasomes" to Liposomes: A Memoir of Alec Bangham, 1921–2010', *The FASEB Journal* 24, pp. 1308–10, DOI: 10.1096/fj.10-0503

'You would never see functioning proteins in this form, as straight chains':
Tandford, Charles (1997), 'How Protein Chemists Learned about the Hydrophobic Factor', *Protein Science* 6, pp. 1358–66, DOI: 10.1002/pro.5560060627

'The story begins with Walter Kauzmann . . .':
McClure, D. S. (2009), 'Walter Kauzmann: 1916–2009', *National Academy*

of Sciences, http://www.nasonline.org/publications/biographical-memoirs/memoir-pdfs/kauzmann-walter.pdf

'Biochemists at the time were starting to ask a lot of questions about proteins':
McClure (2009), op. cit.

'The molecule is nudged to fold into its correct shape':
Ball (2011) – see Chapter One

'Mini power plants inside your cells called mitochondria':
Lane, Nick (2010), 'Why Are Cells Powered by Proton Gradients?', *Nature Education* 3, p. 18, http://www.nature.com/scitable/topicpage/why-are-cells-powered-by-proton-gradients-14373960

'Peter Mitchell had gained a PhD from Cambridge University in 1951':
Slater, E. C. (1994), 'Peter Dennis Mitchell: 29 September 1920–10 April 1992', *Biographical Memoirs of Fellows of the Royal Society*, DOI: 10.1098/rsbm.1994.0040

Biography of Peter Mitchell, http://www.nobelprize.org/nobel_prizes/chemistry/laureates/1978/mitchell-bio.html

The hawala system:
This is well described in this intriguing-sounding US government field guide for its Treasury officials: 'The Hawala Alternative Remittance System and its Role in Money Laundering', http://www.treasury.gov/resource-center/terrorist-illicit-finance/documents/fincen-hawala-rpt.pdf

'This nano-scale hawala system for protons is more formally called the Grotthuss mechanism . . .':
Cukierman, Sam (2006), 'Et Tu, Grotthuss! And Other Unfinished Stories', *Biochimica et Biophysica Acta (BBA) – Bioenergetics* 1757, pp. 876–85, DOI: 10.1016/j.bbabio.2005.12.001

'But this bonding also allow water molecules to form long chains':
Ball, Philip (2007), 'Opinion: The Crucible', *Chemistry World*, http://www.rsc.org/chemistryworld/Issues/2007/February/Opinion Thecrucible.asp

'Mitchell's idea was that the current produced by the flow of protons into the mitochondria . . .':
Mitchell, Peter (1961), 'Coupling of Phosphorylation to Electron and Hydrogen Transfer by a Chemi-osmotic Type of Mechanism', *Nature* 191, pp. 144–8, DOI: 10.1038/191144a0

'Mitchell's hypothesis involved a paradigm shift':
Orgel, Leslie (1999), 'Are You Serious, Dr Mitchell?', *Nature* 402, p. 17, DOI: 10.1038/46903

'Mitchell acknowledged how the scientific community had to test new ideas to destruction':
Mitchell, Peter (1978), Nobel Prize banquet speech, http://www.nobel-prize.org/nobel_prizes/chemistry/laureates/1978/mitchell-speech.html

'Mike Russell of NASA's Jet Propulsion Lab in California proposed an idea in 1989':
Whitfield, John (2009), 'Nascence Man', *Nature* 459, pp. 316–9, DOI: 10.1038/459316a

Russell, Michael et al. (2014), 'The Drive to Life on Wet and Icy Worlds', *Astrobiology* 14, pp. 308–43, DOI: 10.1089/ast.2013.1110

Russell, Michael et al. (2013), 'The Inevitable Journey to Being', *Philosophical Transactions of the Royal Society B* 370, DOI: 10.1098/rstb.2012.0254

General sources:
Ball, Philip (2008), 'Water as an Active Constituent in Cell Biology', *Chemical Reviews* 108, pp. 74–108, http://www.ncbi.nlm.nih.gov/pubmed/18095715

Brindley, Lewis (2009), 'Getting a Look at Water Wires', *Chemistry World*, http://www.rsc.org/chemistryworld/News/2009/July/13070901.asp

Committee on the Limits of Organic Life in Planetary Systems (2007), 'The Limits of Organic Life in Planetary Systems', National Research Council, http://www.nap.edu/catalog/11919/the-limits-of-organic-life-in-planetary-systems

Daniel, Roy et al. (2004), Introduction to Discussion Meeting on the Molecular Basis of Life at the Royal Society – Is Life Possible Without Water?, *Philosophical Transactions of the Royal Society B* 359, p. 1143, DOI: 10.1098/rstb.2004.1507

Editorial (2010), 'Water in Biological Systems', *Physical Chemistry Chemical Physics* 12, pp. 10145–6, DOI: 10.1039/C0CP90061C

Finney, John (2004), 'Water? What's So Special about It?', Philosophical Transactions of the Royal Society B 359, pp. 1145–65, DOI: 10.1098/rstb.2004.1495

Halle, Bertil (2004), 'Protein Hydration Dynamics in Solution: A Critical Survey', *Philosophical Transactions of the Royal Society of London B* 359, pp. 1207–24, DOI: 10.1098/rstb.2004.1499

Homes, Bob (1998), 'Life Is . . ., *New Scientist*, http://www.newscientist.com/article/mg15821385.500-life-is.html

Kim, Seong Keun et al. (2010), 'Water in Biological Systems', *Physical Chemistry Chemical Physics* 12, pp. 10145–6, DOI: 10.1039/c0cp90061c

Lane, Nick (2006), *Power, Sex, Suicide: Mitochondria and the Meaning of Life*, Oxford University Press (Oxford)

Lane, Nick (2010), *Life Ascending: The Ten Great Inventions of Evolution*, Profile (London)

Lilley, Terence (2004), 'So, Why is Water Biologically Important?', *Philosophical Transactions of the Royal Society B* 359, pp. 1321–2, DOI: 10.1098/rstb.2004.1509

Lynden-Bell, Ruth et al. (2010), *Water and Life: The Unique Properties of H_2O*, CRC Press (London)

Meyer, Arthur (1943), 'Water and You', *Scientific American* 168, pp. 114–15, DOI: 10.1038/scientificamerican0343-114

Overbye, Dennis (2011), 'A Romp Through Theories on Origins of Life', *New York Times*, http://www.nytimes.com/2011/02/22/science/22origins.html

Sansom, Mark et al. (2001): 'Biophysics: Water at the Nanoscale,' *Nature* 414, pp. 156–9, DOI: 10.1038/35102651

Schueller, Gretel (1998), 'Stuff of Life', *New Scientist*, http://www.newscientist.com/article/mg15921515.100-stuff-of-life.html

Shortle, David (1999), 'Protein Folding as Seen from Water's Perspective', *Nature Structural & Molecular Biology* 6, pp. 203–205, DOI: 10.1038/6640

Shrope, Mark (2000), 'Power House', *New Scientist*, http://www.newscientist.com/article/mg16622321.500-power-house.html

Szent-Gyorgi, Albert (1937), 'Oxidation, Energy Transfer and Vitamins', Nobel Prize lecture, http://www.nobelprize.org/nobel_prizes/medicine/laureates/1937/szent-gyorgyi-lecture.html

Zaccai, G. (2004), 'The Effect of Water on Protein Dynamics', *Philosophical Transactions of the Royal Society B* 359, pp. 1269–75, DOI: 10.1098/rstb.2004.1503

Chapter Six: Water Footprint

On Leonardo da Vinci and Niccolo Machiaveli's plan to divert the Arno:

The best place to start is the book by historian Roger Masters (1999), *Fortune Is a River: Leonardo Da Vinci and Niccolo Machiavelli's Magnificent Dream to Change the Course of Florentine History*, Plume Books (New York). See also:

Dean, Katrina, 'Keeping Books of Nature: An Introduction to Leonardo da Vinci's Codices Arundel and Leicester', British Library Online Gallery feature, http://www.bl.uk/ttp2/pdf/leonardodean.pdf

Maddox, Brenda (2009), 'The Artist, the Philosopher and the Warrior by Paul Strathern', *New Humanist*, https://newhumanist.org.uk/1973

Pfister, L. et al. (2009), *Leonardo Da Vinci's Water Theory: On the Origin and Fate of Water*, IAHS Press (Oxfordshire)

Strathern, Paul (2009), 'Machiavelli, Leonardo & Borgia: A Fateful Collusion,' *History Today*, http://www.historytoday.com/paul-strathern/machiavelli-leonardo-borgia-fateful-collusion

Victoria and Albert Museum website, 'Leonardo da Vinci: Experience, Experiment, Design', http://www.vam.ac.uk/content/articles/l/leonardo-da-vinci-experience-experiment-design/

Witoszek, Nina (2010), 'Rivers and Humans: The Civilizing Project of Leonardo da Vinci and Niccoló Machiavelli', http://www.cas.uio.no/publications_/transference.php, ISBN: 978-82-996367-7-3

'Our control of water today is so precise that we barely think about it':
Tvedt, Terje (ed.) (2009), 'A History of Water', series from I. B. Tauris and Co Ltd (London), http://www.ibtauris.com/Series/History%20of%20Water.aspx

'In the arid conditions of what is now northern Africa':
Hassan, Fekri, 'Water Management and Early Civilisations', UNESCO International Hydrological Programme, http://webworld.unesco.org/Water/wwap/pccp/cd/pdf/history_future_shared_water_resources/water_management_early.pdf

'The footprint of a product':
Hoekstra, Arjen and Mekonnen, Mesfin (2011), 'The Water Footprint of Humanity', *Proceedings of the National Academy of Sciences* 109, pp. 3232–7, DOI: 10.1073/pnas.1109936109

Hoekstra, Arjen et al. (2011), 'The Green, Blue and Grey Water Footprint of Crops and Derived Crop Products', *Hydrology and Earth System Sciences* 15, pp. 1577–1600, DOI: 10.5194/hess-15-1577-2011

Hoekstra, Arjen and Ercin, A. Ertug (2014), 'Water Footprint Scenarios for 2050: A Global Analysis', *Environment International* 64, pp. 71–82, DOI: 10.1016/j.envint.2013.11.019

General sources:

Capra, Fritjof (2008), *The Science of Leonardo: Inside the Mind of the Great Genius of the Renaissance*, Anchor Books

Delli Priscoli, Jerome (2000), 'Water and Civilization: Using History to Reframe Water Policy Debates and to Build a New Ecological Realism', *Water Policy* 1, pp. 623–36

Gleick, Peter et al. (2010), 'Peak Water Limits to Freshwater Withdrawal and Use', *Proceedings of the National Academy of Sciences* 107, pp. 11155–62, DOI: 10.1073/pnas.1004812107

Langenberg, Heike (2012), 'Hydrology: Complex Water Future', *Nature Geoscience* 5, p. 849, DOI: 10.1038/ngeo1658

PART III: CRYOSPHERE

Chapter Seven: The Ice

'The ice . . . is key to the entire climate, geology and life of Earth':
Bartels-Rausch, Thorsten (2013), 'Ten Things We Need to Know about Ice and Snow', *Nature* 494, pp. 27–9, DOI: 10.1038/494027a

'A quarter of the Earth's ocean area is covered in ice . . .':
Finney, John (2004), 'Ice: The Laboratory in Your Freezer', *Interdisciplinary Science Reviews* 29, pp. 339–51, DOI: 10.1179/030801804225012554

'For the past 2 million years or so, the ice pack has been a permanent fixture':
Newton, Alicia (2010), 'Arctic Ice Through the Ages', *Nature Geoscience* 3, p. 304, DOI: 10.1038/ngeo861

'Movement of the Jakobshavn Isbræ . . .': 193
Joughin, I. (2014), 'Brief Communication: Further Summer Speedup of Jakobshavn Isbræ', *Cryosphere* 8, pp. 209–14, DOI: 10.5194/tc-8-209-2014

'The newest phase . . .':
Falenty, Andrzej and Hansen, Thomas et al. (2014), 'Formation and Properties of Ice XVI Obtained by Emptying a Type sII Clathrate Hydrate', *Nature* 526, pp. 231–4, DOI: 10.1038/nature14014

'A group of geologists proposed that it might be present in cold subconducting slab':
Nina, Craig et al. (2000), 'Possible Presence of High-pressure Ice in Cold Subducting Slabs', *Nature* 408, pp. 844–7, DOI: 10.1038/35048555

General sources:
Franks, Felix (1972): 'Chapter 4: The Properties of Ice', from *Water, a Comprehensive Treatise: The Physics and Physical Chemistry of Water* , *Volume 1*, Plenum Press (New York)

Murray, George (ed.) (1901), *The Antarctic Manual for the Use of the Expedition of 1901*, Royal Geographical Society (London)

National Oceanographic and Atmospheric Administration website: http://oceanservice.noaa.gov/facts/cryosphere.html

National Snow and Ice Data Centre website: http://nsidc.org/cryosphere/allaboutcryosphere.html

Salzmann, Christoph et al. (2011), 'The Polymorphism of Ice: Five Unresolved Questions', *Physical Chemistry Chemical Physics* 13, pp. 18468–80, DOI: 10.1039/C1CP21712G

World Glacier Monitoring Service (2008), 'Antarctica', prepared for the United Nations Environment Programme, http://www.grid.unep.ch/glaciers/pdfs/6_8.pdf

World Glacier Monitoring Service (2008), 'Arctic Islands', prepared for the United Nations Environment Programme, http://www.grid.unep.ch/glaciers/pdfs/6_11.pdf

Chapter Eight: Cape Denison

'The beginnings of this way to see into the past came in 1954 from the Danish geochemist':
Dansgaard, Willi (1954), 'The O-18 Abundance of Fresh Water', *Geochimica et Cosmochimica Acta* 6, pp. 241–60, DOI: 10.1016/0016-7037 (54)90003-4

'By releasing the bubbles trapped at different layers of ice . . .':
Walker, Gabrielle (2004), 'Frozen Time', *Nature* 429, pp. 596–7, DOI: 10.1038/429596a

'Using ice cores to compare the concentration of greenhouse gases':
McManus, Jerry (2004), 'A Great Grand-daddy of Ice Cores', *Nature* 429, pp. 611–12, DOI: 10.1038/429611a

More on ice cores:
'Ice Cores and Climate Change', British Antarctic Survey briefing, http://www.antarctica.ac.uk/press/journalists/resources/science/ice_cores_and_climate_change_briefing-sep10.pdf

'Ice Cores', Natural History Museum website, http://www.nhm.ac.uk/nature-online/environmental-change/measuring-climate-change/ice-cores/

'Arctic Sea Ice News and Analysis', National Snow and Ice Data Centre, http://nsidc.org/arcticseaicenews

'It continued to rise to the point where now it is 28 per cent higher':
Brumfiel, Geoff (2008), 'Ice Cores Reveal Climate Secrets', *Nature*, DOI: 10.1038/news.2008.825

'The Arctic has been warming since the 1980s at around double the global rate':
Larsen, Joan Nymand and Anisimov, Oleg (2014), 'Polar Regions', Chapter 28 from the report prepared by Working Group II for the Fifth Assessment Report of the Intergovernmental Panel on Climate Change, http://ipcc-wg2.gov/AR5/images/uploads/WGIIAR5-Chap28_FGDall.pdf

'It had taken him six years of meticulous planning and \$20m to get his sample':
Christner, Brent et al. (2014), 'A Microbial Ecosystem Beneath the West Antarctic Ice Sheet', *Nature* 512, pp. 310–13, DOI: 10.1038/nature13667

Fox, Douglas (2014), 'Antarctica's Secret Garden', *Nature* 512, pp. 244–6, DOI: 10.1038/512244a

'Wherever there is a hint of liquid water on Earth, there is life':
Rothschild, Lynn and Mancinelli, Rocco (2001), 'Life in Extreme Environments', *Nature* 409, pp. 1092–1101, DOI: 10.1038/35059215

General sources:
Azua-Bustos, Armando et al. (2012), 'Life at the Dry Edge: Microorganisms of the Atacama Desert', *FEBS Letters* 586, pp. 2939–45, DOI: 10.1016/j.febslet.2012.07.025

Ball, Philip (2000), 'The Ice Microbes Cometh', *Nature*, DOI: 10.1038/news000217-1

Barnett, Anna (2009), 'Timeline: Ice Memory', *Nature*, Climate Feedback blog, http://blogs.nature.com/climatefeedback/2009/08/timeline_ice_memory.html

BBC News website (2002), 'Antarctic Ice Shelf Breaks Apart', http://news.bbc.co.uk/1/hi/sci/tech/1880566.stm

Boswell, Evelyn (2014), 'US Expedition Yields First Breakthrough Paper about Life under Antarctic Ice', Montana State University News Service, http://www.montana.edu/news/15002/u-s-expedition-yields-first-breakthrough-paper-about-life-under-antarctic-ice

Jha, Alok (2011), 'Antarctic Mission to Look for Life in Sub-glacial Lake', *Observer*, http://www.theguardian.com/world/2011/oct/15/antarctic-mission-sub-glacial-lake

Jha, Alok, (2012), 'Antarctic Lake Race Sees Scientists Dash for Life's Secrets in Subglacial World', *Guardian*, http://www.theguardian.com/world/2012/feb/14/antarctic-sub-glacial-lake-vostok-life

Jha, Alok (2012), 'Antarctic Lake Research Abandoned', *Guardian*, http://www.theguardian.com/world/2012/dec/27/antarctic-lake-research-abandon/print

Jones, Nicola (2007), 'Polar Research: Buried Treasure', *Nature* 446, pp. 126–8, DOI: 10.1038/446126a

Jones, Nicola (2012), 'Russians Celebrate Vostok Victory', *Nature* 482, p. 287, DOI: 10.1038/482287a

Lake Ellsworth project press statement (December 2012), 'Antarctic Lake Mission Called off', http://www.ellsworth.org.uk/release4_2012.html

Lake Ellsworth project press statement (November 2013), 'Subglacial Lake Ellsworth – 1 year on', http://www.ellsworth.org.uk/november13update.html

Nature Reports: 'Climate Change – Earth's Oldest Ice: A History of Polar Ice-core Research', *Nature*, http://www.nature.com/climate/timeline/polar-ice-cores/index.html

NASA Earth Observatory website (2010): 'World of Change: Collapse of the Larsen-B Ice Shelf', http://earthobservatory.nasa.gov/Features/WorldOfChange/larsenb.php

Newton, Alicia (2007), 'Cryosphere: The Big Melt', *Nature Geoscience* 1, DOI: 10.1038/ngeo.2007.31

Oerlemans, Johannes (2009), 'A World Without Ice', *Nature* 462, pp. 572–3, DOI: 10.1038/462572a

Peplow, Mark (2004), 'Glacial Lake Hides Bacteria', *Nature*, DOI: 10.1038/news040712-6,

Peplow, Mark (2004), 'Ice Core Reveals Gentle Start to Last Ice Age', *Nature*, DOI: 10.1038/news040906-10,

Qui, Jane (2009), 'China Builds Inland Antarctic Base', *Nature*, DOI: 10.1038/457134a

Sample, Ian (2012), 'Search for Life Begins in Lake Entombed under Antarctic Ice', *Guardian*, http://www.theguardian.com/science/2012/dec/02/search-life-lake-antarctic-ice

Scharf, Caleb (2012), 'Lake Vostok is (Almost) Breached After 20 Million Years', *Life*, Unbounded blog, http://blogs.scientificamerican.com/life-unbounded/2012/02/06/lake-vostok-is-almost-breached-after-20-million-years/

Scharf, Caleb (2013), 'Lake Vostok Water Ice Has Been Obtained', *Life*, Unbounded blog, http://blogs.scientificamerican.com/life-unbounded/2013/01/13/lake-vostok-water-ice-has-been-obtained/

Schiermeier, Quirin (2012), 'Hunt for Life under Antarctic Ice Heats up', *Nature* 491, pp. 506–7, DOI: 10.1038/491506a

Schiermeier, Quirin (2012), 'Life Abounds in Antarctic Lake Sealed under Ice', *Nature*, DOI: 10.1038/nature.2012.11884

Schiermeier, Quirin (2013), 'Life Discovered under Ice in Antarctic Lake', *Nature*, DOI: 10.1038/nature.2013.12405

Schiermeier, Quirin (2013), 'Claims of Lake Vostok Fish Get Frosty Response', *Nature*, DOI: 10.1038/nature.2013.13364

PART IV: SPACE

Chapter Nine: The Moon

More on the differences between the atmospheres of the solar system's inner planets:
Kasting, James et al. (1988), 'How Climate Evolved on the Terrestrial Planets', *Scientific American* 258, pp. 90–7, DOI: 10.1038/scientificamerican 0288-90

The LCROSS mission's published results and details of operation:
Colaprete, Anthony et al. (2010), 'Detection of Water in the LCROSS Ejecta Plume', *Science* 330, pp. 463–468, DOI: 10.1126/science.1186986

Gladstone, Randall et al, (2010), 'LRO-LAMP Observations of the LCROSS Impact Plume', *Science* 330, pp. 472–6, DOI: 10.1126/science.1186474

Schultz, Peter et al. (2010), 'The LCROSS Cratering Experiment,' *Science* 330, pp. 468–72, DOI: 10.1126/science.1187454

Heldmann, Jennifer et al. (2012), 'LCROSS (Lunar Crater Observation and Sensing Satellite) Observation Campaign: Strategies, Implementation, and Lessons Learned,' *Space Science Reviews* 167, pp. 93–140, DOI: 10.1007/s11214-011-9759-y

News stories and statements from NASA about the initial LCROSS discoveries:
NASA science news (2009), 'Water Molecules Found on the Moon', http://science.nasa.gov/science-news/science-at-nasa/2009/24sep_moonwater/

NASA LCROSS media kit, http://web.archive.org/web/20091027111849/http://www.nasa.gov/pdf/360020main_LRO_LCROSS_presskit2.pdf

NASA press release (2009), 'LCROSS Impact Data Indicates Water on Moon', http://www.nasa.gov/mission_pages/LCROSS/main/prelim_water_results.html#.VJmfQAgI

'Scientists had largely come to the conclusion that the Moon was dry and airless':
Crott, Arlin (2012), 'Water on The Moon, I. Historical Overview', *Astronomical Review*, http://astroreview.com/issue/2012/article/water-on-the-moon-i-historical-overview

Harold Urey biography on the Nobel Prize website: http://www.nobelprize.org/nobel_prizes/chemistry/laureates/1934/urey-bio.html

Chapter Ten: Mars

The dramatic landing of Curiosity on Mars was recounted in several newspaper articles and NASA video including:
McKie, Robin (2012), 'NASA Counts Down the Hours to its Latest Mission: Is There Life on Mars?', *Observer*, http://www.theguardian.com/science/2012/jul/14/nasa-mission-to-mars

Sample, Ian (2012), 'Mars Curiosity Rover: NASA Scientists Brace for "Seven Minutes of Terror"', *Guardian*, http://www.theguardian.com/science/2012/aug/03/mars-explorer-rover-seven-minutes

'The Challenge of Getting to Mars: Curiosity's Seven Minutes of Terror', NASA Jet Propulsion Laboratory Video Transcript, http://mars.jpl.nasa.gov/multimedia/videos/movies/curiosity20120622/curiosity20120622.pdf

Grossman, Lisa (2012), 'NASA Jubilant as Curiosity Rover Lands safely on Mars', *New Scientist*, http://www.newscientist.com/article/dn22141-nasa-jubilant-as-curiosity-rover-lands-safely-on-mars.html#.U-mhyoBdVy8

'Ever since Galileo spotted Mars through his telescope four centuries ago':
Titus, Timothy (2014), 'Water, Water Everywhere', *Nature* 428, pp. 610–11, DOI: 10.1038/nature02482

Clark, Stuart (2012), 'Water on Mars Has a Long History', *Guardian*, http://www.theguardian.com/science/across-the-universe/2012/oct/01/water-mars-history

General sources:
Baker, Victor (2007), 'Water Cycling on Mars', *Nature* 446, pp. 150–1, DOI: 10.1038/446150b

Bell, Jim (2006), 'The Red Planet's Watery Past', *Scientific American* 295, pp. 62–9, DOI: 10.1038/scientificamerican1206-62

Clark, Stuart (2012), 'NASA's Messenger Finds Water Ice and Organic Molecules on Mercury', *Guardian*, http://www.theguardian.com/science/across-the-universe/2012/nov/30/nasa-messenger-water-ice-organic-molecules-mercury

Jha, Alok (2013), 'NASA's Curiosity Rover Finds Water in Martian Soil', *Guardian*, http://www.theguardian.com/science/2013/sep/26/nasa-curiosity-rover-mars-soil-water

Kerr, Richard (2013), 'New Results Send Mars Rover on a Quest for Ancient Life', *Science*, http://news.sciencemag.org/chemistry/2013/12/new-results-send-mars-rover-quest-ancient-life

Lakdawalla, Emily (2013), 'Curiosity's First Year on Mars: Where's the Science?', The Planetary Society, http://www.planetary.org/blogs/emily-lakdawalla/2013/08071156-curiosity-1yearonmars-science.html

McKee, Maggie (2013), 'NASA's 2020 Mars Rover Would Search for Signs of Past life,' *Nature*, DOI: 10.1038/nature.2013.13365

Planetary Science Institute website (2014), 'Curiosity Rover Has Seen Much in First Two Years on Mars', http://www.psi.edu/news/curiosity2years

Sagan, Carl (1994), 'The Search for Extraterrestrial Life', *Scientific American* 271, pp. 92–9, DOI: 10.1038/scientificamerican1094-92

Witze, Alexandra (2013), 'Mars Rover Finds Evidence of Ancient Habitability', *Nature*, DOI: 10.1038/nature.2013.12597

Witze, Alexandra (2014), 'Old Mars Rover Finds Signs of Ancient Water,' *Nature*, DOI: 10.1038/nature.2014.14569

Chapter Eleven: Solar System Moons

'Observations from telescopes on the ground in the 1950s':
Greenberg, Ralph (2002), 'An Ocean on Europa – The History of an Idea', https://www.math.washington.edu/~greenber/EuropaHistory.html

'The idea that these moons might contain liquid water began in 1971 with John S. Lewis':
Lewis, John (1971), 'Satellites of the Outer Planets: Their Physical and Chemical Nature', *Icarus* 15, pp. 174–185, DOI: 10.1016/0019-1035 (71)90072-8

Discovery of 'Cold Faithful' on Enceladus:
Spencer, John et al. (2006), 'Cassini Encounters Enceladus: Background and the Discovery of a South Polar Hotspot', *Science* 311, pp. 1401–5, DOI: 10.1126/science.1121661

Porco, Carolyn et al. (2006), 'Cassini Observes the Active South Pole of Enceladus', *Science* 311, pp. 1393–1401, DOI: 10.1126/science.1123013

Kieffer, Susan and Jakosky, Bruce (2008), 'Enceladus – Oasis or Ice Ball?', *Science* 320, pp. 1432–3, DOI: 10.1126/science.1159702

Discovery of oceans on Enceladus:
Iess, Luciano et al. (2014), 'The Gravity Field and Interior Structure of Enceladus', *Science* 344, pp. 78–80, DOI: 10.1126/science.1250551

Witze, Alexandra (2014), 'Icy Enceladus Hides a Watery Ocean', *Nature*, DOI: 10.1038/nature.2014.14985

Chris McKay's proposal for a life form on Titan:
McKay, Chris et al. (2005), 'Possibilities for Methanogenic Life in Liquid Methane on the Surface of Titan', *Icarus* 178, pp. 274–6, DOI: 10.1016/j. icarus.2005.05.018

General sources:
Baker, Joanne (2006), 'Tiger, Tiger, Burning Bright', *Science* 311, p. 1388, DOI: 10.1126/science.311.5766.1388

Ball, Philip (2005), 'Titanic Life May Bloom Without Water', *Nature*, DOI: 10.1038/news050131-2

McKay, Chris et al. (2014), 'Astrobiology', *Encyclopaedia of the Solar System*, DOI: 10.1016/B978-0-12-415845-0.00010-4

Moskowitz, Clara (2014), 'Europa's Water Geysers Entice Scientists to Send a Probe – But Can NASA Do It on the Cheap?', *Scientific American*, http://www.scientificamerican.com/article/nasa-mission-to-europa/

Mottl, Michael et al. (2007), 'Water and Astrobiology', *Chemie der Erde – Geochemistry* 67, pp. 253–82, DOI: 10.1016/j.chemer.2007.09.002

Reynolds, Ray and McKay, Chris et al. (1987), 'Europa, Tidally Heated Oceans, and Habitable Zones Around Giant Planets', *Advances in Space Research* 7, pp. 125–32, DOI: 0.1016/0273-1177(87)90364-4

Witze, Alexandra (2013), 'Hubble Spots Water Spurting from Europa', *Nature*, DOI: 10.1038/nature.2013.14357

Chapter Twelve: Beyond the Solar System

Review paper on extending the range of options for potential stars that could have habitable planets:
Tarter, Jill et al. (2007), 'A Reappraisal of the Habitability of Planets Around M Dwarf Stars', *Astrobiology* 7, pp. 30–65, DOI: 10.1089/ast.2006.0124

How the casting net for habitability keeps getting wider:
Rekola, Rami (2009), 'Life and Habitable Zones in the Universe', *Planetary and Space Science* 57, pp. 430–33, DOI: 10.1016/j.pss.2008.09.011

Discovery of Kepler-186f:
Quintana, Elisa et al. (2014), 'An Earth-Sized Planet in the Habitable Zone of a Cool Star', *Science* 344, pp. 277–280, DOI: 10.1126/science.1249403

Quintana, Elisa (2014), 'Kepler 186f: First Earth-sized Planet Orbiting in Habitable Zone of Another Star', SETI Institute website, http://www.seti.org/seti-institute/kepler-186f-first-earth-sized-planet-orbiting-in-habitable-zone-of-another-star

'Astronomers [. . .] measured the light coming from the planet and its star':
Powell, Devin (2014), 'First Distant Planet to be Seen in Colour Is Blue', *Nature*, DOI: 10.1038/nature.2013.13376

General sources:
Fritz, Jeorg (2014), 'Earth-like Habitats in Planetary Systems', *Planetary and Space Science* 98, pp. 254–267, DOI: 10.1016/j.pss.2014.03.003

Jha, Alok (2011), 'Two New Planets Are Most Earth-like Ever Seen – But Hot as Hell', Guardian, http://www.theguardian.com/science/2011/dec/20/planets-earth-like-exoplanet-solar-system

Jha, Alok (2013), 'Two Billion Planets in our Galaxy May Be Suitable for Life', *Guardian*, http://www.theguardian.com/science/2013/nov/04/planets-galaxy-life-kepler

McKay, Chris (2014), 'Requirements and Limits for Life in the Context of Exoplanets', *Proceedings of the National Academy of Sciences* 111, pp. 12628–33, DOI: 10.1073/pnas.1304212111

Chapter Thirteen: Alternatives to Water

'There is a famous book . . .':
Sagan, Carl (ed.) (1973), *Communication with Extraterrestrial Intelligence (CETI)*, MIT Press (Cambridge, Massachusetts)

'Chemical parochialism':
Bains, William, 'Many Chemistries Could Be Used to Build Living Systems', *Astrobiology* 4, p. 137–67, DOI: 10.1089/153110704323175124

A major review convened by the US National Academy of Sciences on how to look for life in space, containing many useful descriptions of what life might be like with alternative chemistries:

Committee on the Limits of Organic Life in Planetary Systems (2007), 'The Limits of Organic Life in Planetary Systems', National Research Council, http://www.nap.edu/catalog/11919/the-limits-of-organic-life-in-planetary-systems

'A protein-based organism . . .':

Cleland, Carol and Chyba, Christopher (2002), 'Defining "Life"', *Origins of Life and Evolution of the Biosphere* 32, pp. 387–93, DOI: 10.1023/A:1020503324273

'Even so, says chemist Steven Benner . . .':

Benner, Steven et al. (2004): 'Is There a Common Chemical Model for Life in the Universe?', *Current Opinion in Chemical Biology* 8, pp. 672–89, DOI: 10.1016/j.cbpa.2004.10.003

'Suspected to exist in underground regions of Mars . . .':

Ball, Philip (2005), 'Water and Life: Seeking the Solution', *Nature* 436, pp. 1084–5, DOI: 10.1038/4361084a

'Biochemical reactions in this pre-complex-life world . . .':

Wolfenden, Richard et al. (2010), 'Impact of Temperature on the Time Required for the Establishment of Primordial Biochemistry, and for the Evolution of Enzymes', *Proceedings of the National Academy of Sciences* 107, pp. 22102–5, DOI: 10.1073/pnas.1013647107

Ball, Philip (2010), 'Some Like it Hot', *Nature*, DOI: 10.1038/news.2010.590

General source:

Koshland, Daniel (2002), 'The Seven Pillars of Life', *Science* 295, pp. 2215–6, DOI: 10.1126/science.1068489

Epilogue

Krauss, Lawrence and Starkman, Glenn (2002), 'The Fate of Life in the Universe', *Scientific American* 12, pp. 50–57, DOI: 10.1038/scientificamerican1002-50sp

Index